T0239448

Lecture Notes in Computer Science   8882

Commenced Publication in 1973
Founding and Former Series Editors:
Gerhard Goos, Juris Hartmanis, and Jan van Leeuwen

Editorial Board

David Hutchison
 Lancaster University, Lancaster, UK
Takeo Kanade
 Carnegie Mellon University, Pittsburgh, PA, USA
Josef Kittler
 University of Surrey, Guildford, UK
Jon M. Kleinberg
 Cornell University, Ithaca, NY, USA
Friedemann Mattern
 ETH Zürich, Zürich, Switzerland
John C. Mitchell
 Stanford University, Stanford, CA, USA
Moni Naor
 Weizmann Institute of Science, Rehovot, Israel
C. Pandu Rangan
 Indian Institute of Technology, Madras, India
Bernhard Steffen
 TU Dortmund University, Dortmund, Germany
Demetri Terzopoulos
 University of California, Los Angeles, CA, USA
Doug Tygar
 University of California, Berkeley, CA, USA
Gerhard Weikum
 Max Planck Institute for Informatics, Saarbruecken, Germany

More information about this series at http://www.springer.com/series/7407

Anthony Bonato · Fan Chung Graham
Paweł Prałat (Eds.)

Algorithms and Models for the Web Graph

11th International Workshop, WAW 2014
Beijing, China, December 17–18, 2014
Proceedings

 Springer

Editors
Anthony Bonato
Ryerson University
Toronto, Ontario
Canada

Paweł Prałat
Ryerson University
Toronto, Ontario
Canada

Fan Chung Graham
University of California San Diego
La Jolla, California
USA

ISSN 0302-9743
ISBN 978-3-319-13122-1
DOI 10.1007/978-3-319-13123-8

ISSN 1611-3349 (electronic)
ISBN 978-3-319-13123-8 (eBook)

Library of Congress Control Number: 2014954619

LNCS Sublibrary: SL1 – Theoretical Computer Science and General Issues

Springer Cham Heidelberg New York Dordrecht London

© Springer International Publishing Switzerland 2014

This work is subject to copyright. All rights are reserved by the Publisher, whether the whole or part of the material is concerned, specifically the rights of translation, reprinting, reuse of illustrations, recitation, broadcasting, reproduction on microfilms or in any other physical way, and transmission or information storage and retrieval, electronic adaptation, computer software, or by similar or dissimilar methodology now known or hereafter developed. Exempted from this legal reservation are brief excerpts in connection with reviews or scholarly analysis or material supplied specifically for the purpose of being entered and executed on a computer system, for exclusive use by the purchaser of the work. Duplication of this publication or parts thereof is permitted only under the provisions of the Copyright Law of the Publisher's location, in its current version, and permission for use must always be obtained from Springer. Permissions for use may be obtained through RightsLink at the Copyright Clearance Center. Violations are liable to prosecution under the respective Copyright Law.

The use of general descriptive names, registered names, trademarks, service marks, etc. in this publication does not imply, even in the absence of a specific statement, that such names are exempt from the relevant protective laws and regulations and therefore free for general use.

While the advice and information in this book are believed to be true and accurate at the date of publication, neither the authors nor the editors nor the publisher can accept any legal responsibility for any errors or omissions that may be made. The publisher makes no warranty, express or implied, with respect to the material contained herein.

Printed on acid-free paper

Springer is part of Springer Science+Business Media (www.springer.com)

Preface

The 11th Workshop on Algorithms and Models for the Web Graph (WAW 2014) took place at the Academy of Mathematics and Systems Science in Beijing, China, during December 17–18, 2014. This is an annual meeting, which is traditionally colocated with another, related, conference. WAW 2014 was colocated with the 10th Conference on Web and Internet Economics (WINE 2014). Colocation of the workshop and conference provided opportunities for researchers in two different but interrelated areas to interact and to exchange research ideas. It was an effective venue for the dissemination of new results and for fostering research collaboration.

The World Wide Web has become a part of our everyday life, and information retrieval and data mining on the Web are now of enormous practical interest. The algorithms supporting these activities combine the view of the Web as a text repository and as a graph, induced in various ways by links among pages, hosts, and users. The aim of the workshop was to further the understanding of graphs that arise from the Web and various user activities on the Web, and stimulate the development of high-performance algorithms and applications that exploit these graphs. The workshop gathered the researchers who are working on graph-theoretic and algorithmic aspects of related complex networks, including social networks, citation networks, biological networks, molecular networks, and other networks arising from the Internet.

This volume contains the papers presented during the workshop. Each submission was reviewed by the Programme Committee members. Papers were submitted and reviewed using the EasyChair online system. The committee members decided to accept 12 papers.

December 2014

Anthony Bonato
Fan Chung Graham
Paweł Prałat

Organization

General Chairs

Andrei Z. Broder   Google Research, USA
Fan Chung Graham   University of California, San Diego, USA

Organizing Committee

Anthony Bonato   Ryerson University, Canada
Fan Chung Graham   University of California, San Diego, USA
Paweł Prałat   Ryerson University, Canada

Program Committee

Konstantin Avratchenkov   Inria, France
Ayse Bener   Ryerson University, Canada
Paolo Boldi   University of Milano, Italy
Anthony Bonato   Ryerson University, Canada
Milan Bradonjic   Bell Laboratories, USA
Fan Chung Graham   University of California, San Diego, USA
Collin Cooper   King's College London, UK
Artur Czumaj   University of Warwick, UK
Andrzej Dudek   Western Michigan University, USA
Alan Frieze   Carnegie Mellon University, USA
David Gleich   Purdue University, USA
Adam Henry   University of Arizona, USA
Jeannette Janssen   Dalhousie University, Canada
Evangelos Kranakis   Carleton University, Canada
Ravi Kumar   Google, USA
Stefano Leonardi   Sapienza University of Rome, Italy
Marek Lipczak   Dalhousie University, Canada
Nelly Litvak   University of Twente, The Netherlands
Linyuan Lu   University of South Carolina, USA
Michael Mahoney   UC Berkeley, USA
Oliver Mason   NUI Maynooth, Ireland
Dieter Mitsche   Université de Nice Sophia-Antipolis, France
Peter Morters   University of Bath, UK
Tobias Mueller   Utrecht University, The Netherlands

Mariana Olvera-Cravioto	Columbia University, USA
JP Onnela	Harvard University, USA
Liudmila Ostroumova	Yandex, Russia
Pan Peng	TU Dortmund, Germany
Paweł Prałat	Ryerson University, Canada
Stephen Young	University of Louisville, USA

Sponsoring Institutions

Google
Internet Mathematics
Microsoft Research New England
Ryerson University

Contents

Clustering and the Hyperbolic Geometry of Complex Networks.   1
  Elisabetta Candellero and Nikolaos Fountoulakis

Burning a Graph as a Model of Social Contagion.   13
  Anthony Bonato, Jeannette Janssen, and Elham Roshanbin

Personalized PageRank with Node-Dependent Restart.   23
  Konstantin Avrachenkov, Remco van der Hofstad, and Marina Sokol

Efficient Computation of the Weighted Clustering Coefficient.   34
  Silvio Lattanzi and Stefano Leonardi

Global Clustering Coefficient in Scale-Free Networks.   47
  Liudmila Ostroumova Prokhorenkova and Egor Samosvat

Efficient Primal-Dual Graph Algorithms for MapReduce.   59
  Bahman Bahmani, Ashish Goel, and Kamesh Munagala

Computing Diffusion State Distance Using Green's Function
and Heat Kernel on Graphs   79
  Edward Boehnlein, Peter Chin, Amit Sinha, and Linyuan Lu

Relational Topic Factorization for Link Prediction in Document Networks. . . .   96
  Wei Zhang, Jiankou Li, and Xi Yong

Firefighting as a Game   108
  Carme Àlvarez, Maria J. Blesa, and Hendrik Molter

PageRank in Scale-Free Random Graphs.   120
  Ningyuan Chen, Nelly Litvak, and Mariana Olvera-Cravioto

Modelling of Trends in Twitter Using Retweet Graph Dynamics   132
  Marijn ten Thij, Tanneke Ouboter, Daniël Worm, Nelly Litvak,
  Hans van den Berg, and Sandjai Bhulai

LiveRank: How to Refresh Old Crawls   148
  The Dang Huynh, Fabien Mathieu, and Laurent Viennot

Author Index   161

Clustering and the Hyperbolic Geometry of Complex Networks

Elisabetta Candellero[1] and Nikolaos Fountoulakis[2]([⊠])

[1] Department of Statistics, University of Warwick Coventry,
Coventry CV4 7AL, UK
elisabetta.candellero@gmail.com
[2] School of Mathematics, University of Birmingham,
Edgbaston B15 2TT, UK
n.fountoulakis@bham.ac.uk

Abstract. Clustering is a fundamental property of complex networks and it is the mathematical expression of a ubiquitous phenomenon that arises in various types of self-organized networks such as biological networks, computer networks or social networks. In this paper, we consider what is called the *global clustering coefficient* of random graphs on the hyperbolic plane. This model of random graphs was proposed recently by Krioukov et al. [22] as a mathematical model of complex networks, implementing the assumption that hyperbolic geometry underlies the structure of these networks. We do a rigorous analysis of clustering and characterize the global clustering coefficient in terms of the parameters of the model. We show how the global clustering coefficient can be tuned by these parameters, giving an explicit formula.

1 Introduction

The theory of complex networks was developed during the last 15 years mainly as a unifying mathematical framework for modeling a variety of networks such as biological networks or large computer networks among which is the Internet, the World Wide Web as well as social networks that have been recently developed over these platforms. A number of mathematical models have emerged whose aim is to describe fundamental characteristics of these networks as these have been described by experimental evidence – see for example [1]. Among the most influential models was the Watts-Strogatz model of small worlds [30] and the Barabási-Albert model [3], that is also known as the preferential attachment model. The main typical characteristics of these networks have to do with the distribution of the degrees (e.g., power-law distribution), the existence of clustering as well as the typical distances between vertices (e.g., the small world effect).

Loosely speaking, the notion of a complex network refers to a class of large networks which exhibit the following characteristics:

Nikolaos Fountoulakis: This research has been supported by a Marie Curie Career Integration Grant PCIG09-GA2011-293619.

© Springer International Publishing Switzerland 2014
A. Bonato et al. (Eds.): WAW 2014, LNCS 8882, pp. 1–12, 2014.
DOI: 10.1007/978-3-319-13123-8_1

1. they are *sparse*, that is, the number of their edges is proportional to the number of nodes;
2. they exhibit the *small world phenomenon*: most pairs of vertices which belong to the same component are within a short distance from each other;
3. *clustering*: two nodes of the network that have a common neighbour are somewhat more likely to be connected with each other;
4. the tail of their degree distribution follows a *power law*: experimental evidence (see [1]) indicates that many networks that emerge in applications follow power law degree distribution with exponent between 2 and 3.

The books of Chung and Lu [13] and of Dorogovtsev [15] are excellent references for a detailed discussion of these properties.

The models that we described above as well as other known models, such as the Chung-Lu model (defined by Chung and Lu [11], [12]) fail to capture *all* the above features simultaneously or if they do so, they do it in a way that is difficult to tune these features independently. For example, the Barabasi-Albert model (when suitably parametrized) exhibits a power law degree distribution with exponent between 2 and 3, and average distance of order $O(\log \log N)$, but it is locally tree-like around a typical vertex (cf. [8], [16]). On the other hand, the Watts-Strogatz model, although it exhibits clustering and small distances between the vertices, has degree distribution that decays exponentially [4].

The notion of clustering formalizes the property that two nodes of a network that share a neighbor (for example two individuals that have a common friend) are more likely to be joined by an edge (that is, to be friends of each other). In the context of social networks, sociologists have explained this phenomenon through the notion of *homophily*, which refers to the tendency of individuals to be related with similar individuals, e.g. having similar socioeconomic background or similar educational background. There have been numerous attempts to define models where clustering is present – see for example the work of Coupechoux and Lelarge [14] or that of Bollobás, Janson and Riordan [9] where this is combined with the general notion of inhomogeneity. In that context, clustering is *planted* in a sparse random graph. Also, it is even more rare to quantify clustering precisely (as for example in random intersection graphs [5]). This is the case as the presence of clustering is the outcome of heavy dependencies between the edges of the random graphs and, in general, these are not easy to handle.

However, clustering is naturally present on random graphs that are created on a metric space, as is the case of a random geometric graph on the Euclidean plane. The theory of random geometric graphs was initiated by Gilbert [18] already in 1961 and started taking its present form later by Hafner [20]. In its standard form a geometric random graph is created as follows: N points are sampled within a subset of $\mathbb{R}^d$ following a particular distribution (most usually this is the uniform distribution or the distribution of the point-set of a Poisson point process) and any two of them are joined when their Euclidean distance is smaller than some threshold value which, in general, is a function of N. During the last two decades, this kind of random graphs was studied in depth by several researchers – see the monograph of Penrose [29] and the references therein. Numerous typical properties of such random graphs have been investigated, such as the chromatic number [24], Hamiltonicity [2] etc.

There is no particular reason why a random geometric graph on a Euclidean space would be intrinsically associated with the formation of a complex network. Real-world networks consist of heterogeneous nodes, which can be classified into groups. In turn, these groups can be classified into larger groups which belong to bigger subgroups and so on. For example, if we consider the network of citations, whose set of nodes is the set of research papers and there is a link from one paper to another if one cites the other, there is a natural classification of the nodes according to the scientific fields each paper belongs to (see for example [10]). In the case of the network of web pages, a similar classification can be considered in terms of the similarity between two web pages: the more similar two web pages are, the more likely it is that there exists a hyperlink between them [25].

This classification can be approximated by tree-like structures representing the hidden hierarchy of the network. The tree-likeness suggests the hypothesis that the geometry of this hierarchy is *hyperbolic*. One of the basic features of a hyperbolic space is that the volume growth is exponential which is also the case, for example, when one considers a k-ary tree, that is, a rooted tree where every vertex has k children. Let us consider for example the Poincaré unit disc model (which we will discuss in more detail in the next section). If we place the root of an infinite k-ary tree at the centre of the disc, then the hyperbolic metric provides the necessary room to embed the tree into the disc so that every edge has unit length in the embedding.

Recently Krioukov et al. [22] introduced a model which implements this idea. In this model, a random network is created on the hyperbolic plane (we will see the detailed definition shortly). In particular, Krioukov et al. [22] determined the degree distribution for *large* degrees showing that it is *scale free* and its tail follows a power law, whose exponent is determined by some of the parameters of the model. Furthermore, they consider the clustering properties of the resulting random network. A numerical approach in [22] suggests that the (local) clustering coefficient[1] is positive and it is determined by one of the parameters of the model. In fact, as we will discuss in Section 2, this model corresponds to the sparse regime of random geometric graphs on the hyperbolic plane and hence is of independent interest within the theory of random geometric graphs.

This paper investigates *rigorously* the presence of clustering in this model, through the notion of the *clustering coefficient*. Our first contribution is that we manage to determine exactly the value of the clustering coefficient as a function of the parameters of the model. More importantly, our results imply that in fact the exponent of the power law, the density of the random graph and the amount of clustering can be tuned *independently* of each other, through the parameters of the random graph. Furthermore, we should point out that the clustering coefficient we consider is the so-called *global clustering coefficient*. Its calculation involves tight concentration bounds on the number of triangles in the random graph. Hence, our analysis initiates an approach to the small subgraph counting

---

[1] This is defined as the average density of the neighbourhoods of the vertices.

problem in these random graphs, which is among the central problems in the general theory of random graphs [7],[21] and of random geometric graphs [29].

1.1   Random Geometric Graphs on the Hyperbolic Plane

The most common representations of the hyperbolic plane are the upper-half plane representation $\{z \in \mathbb{C} \ : \ \Im z > 0\}$ as well as the Poincaré unit disc which is simply the open disc of radius one, that is, $\{(u, v) \in \mathbb{R}^2 \ : \ 1 - u^2 - v^2 > 0\}$. Both spaces are equipped with the hyperbolic metric; in the former case this is $\frac{1}{(\zeta y)^2} dy^2$ whereas in the latter this is $\frac{4}{\zeta^2} \frac{du^2 + dv^2}{(1-u^2-v^2)^2}$, where ζ is some positive real number. It can be shown that the (Gaussian) curvature in both cases is equal to $-\zeta^2$ and the two spaces are isometric, i.e., there is a bijection between the two spaces that preserves (hyperbolic) distances. In fact, there are more representations of the 2-dimensional hyperbolic space of curvature $-\zeta^2$ which are isometrically equivalent to the above two. We will denote by $\mathbb{H}_\zeta^2$ the class of these spaces.

In this paper, following the definitions in [22], we shall be using the *native* representation of $\mathbb{H}_\zeta^2$. Here, the ground space of $\mathbb{H}_\zeta^2$ is $\mathbb{R}^2$ and every point $x \in \mathbb{R}^2$ whose polar coordinates are (r, θ) has hyperbolic distance from the origin equal to r. More precisely, the native representation can be viewed as a mapping of the Poincaré unit disc to $\mathbb{R}^2$, where the origin of the unit disc is mapped to the origin of $\mathbb{R}^2$ and every point v in the Poincaré disc is mapped to a point $v' \in \mathbb{R}^2$, where $v' = (r, \theta)$ in polar coordinates: r is the hyperbolic distance of v from the origin of the Poincaré disc and θ is its angle.

An elementary but tedious calculation can show that a circle of radius r around the origin has length equal to $\frac{2\pi}{\zeta} \sinh \zeta r$ and area equal to $\frac{2\pi}{\zeta^2}(\cosh \zeta r - 1)$.

We are now ready to give the definitions of the two basic models introduced in [22]. Consider the native representation of the hyperbolic plane of curvature $K = -\zeta^2$, for some $\zeta > 0$. For some constant $\nu > 0$, we let $N = \nu e^{\zeta R/2}$ – thus R grows logarithmically as a function of N. We shall explain the role of ν shortly. We create a random graph by selecting randomly and independently N points from the disc of radius R centered at the origin O, which we denote by $\mathcal{D}_R$.

The distribution of these points is as follows. Assume that a random point u has polar coordinates (r, θ). The angle θ is uniformly distributed in $(0, 2\pi]$ and the probability density function of r, which we denote by $\rho_N(r)$, is determined by a parameter $\alpha > 0$ and is equal to

$$\rho_N(r) = \begin{cases} \alpha \frac{\sinh \alpha r}{\cosh \alpha R - 1}, & \text{if } 0 \le r \le R \\ 0, & \text{otherwise} \end{cases} \tag{1}$$

Note that when $\alpha = \zeta$, this is simply the uniform distribution.

An alternative way to define this distribution is as follows. Consider $\mathbb{H}_\alpha^2$ and the Poincaré representation of it. Let O' be the centre of the disc. Consider the disc $\mathcal{D}_R'$ of radius R around O' and select N points within $\mathcal{D}_R'$ uniformly at random. Subsequently, the selected points are projected onto $\mathcal{D}_R$ preserving their polar coordinates. The projections of these points, which we will be denoting by V_N, will be the vertex set of the random graph. We will be also treating the vertices as points in the hyperbolic space indistinguishably.

Note that the curvature in this case determines the rate of growth of the space. Hence, when $\alpha < \zeta$, the N points are distributed on a disc (namely $\mathcal{D}'_R$) which has smaller area compared to $\mathcal{D}_R$. This naturally increases the density of those points that are located closer to the origin. Similarly, when $\alpha > \zeta$ the area of the disc $\mathcal{D}'_R$ is larger than that of $\mathcal{D}_R$, and most of the N points are significantly more likely to be located near the boundary of $\mathcal{D}'_R$, due to the exponential growth of the volume.

Given the set V_N on $\mathcal{D}_R$ we define the following two models of random graphs.

1. *The disc model*: this model is the most commonly studied in the theory of random geometric graphs on Euclidean spaces. We join two vertices if they are within (hyperbolic) distance R from each other. Figure 1 illustrates a disc of radius R around a vertex $v \in \mathcal{D}_R$.

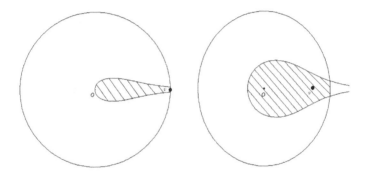

Fig. 1. The disc of radius R around v in $\mathcal{D}_R$

2. *The binomial model*: we join any two distinct vertices u, v with probability

$$p_{u,v} = \frac{1}{\exp\left(\beta \frac{\zeta}{2}(d(u,v) - R)\right) + 1},$$

independently of every other pair, where $\beta > 0$ is fixed and $d(u,v)$ is the hyperbolic distance between u and v. We denote the resulting random graph by $\mathcal{G}(N; \zeta, \alpha, \beta, \nu)$.

The binomial model is in some sense a *soft* version of the disc model. In the latter, two vertices become adjacent if and only if their hyperbolic distance is at most R. This is *approximately* the case in the former model. If $d(u,v) = (1+\delta)R$, where $\delta > 0$ is some small constant, then $p_{u,v} \to 0$, whereas if $d(u,v) = (1-\delta)R$, then $p_{u,v} \to 1$, as $N \to \infty$.

Also, the disc model can be viewed as a limiting case of the binomial model as $\beta \to \infty$. Assume that the positions of the vertices in $\mathcal{D}_R$ have been realized. If $u, v \in V_N$ are such that $d(u, v) < R$, then when $\beta \to \infty$ the probability that u and v are adjacent tends to 1; however, if $d(u, v) > R$, then this probability converges to 0 as β grows.

Müller [26] has shown that the disc model is in fact determined by the ratio ζ/α. In that case, one may set $\zeta = 1$ and keep only α as the parameter of the model.

Krioukov et al. [22] provide an argument which indicates that in both models the degree distribution has a power law tail with exponent that is equal to $2\alpha/\zeta + 1$. Hence, when $0 < \zeta/\alpha < 2$, any exponent greater than 2 can be realised. This has been shown rigorously by Gugelmann et al. [19], for the disc model, and by the second author [17], for the binomial model. In the latter case, the average degree of a vertex depends on all four parameters of the model. For the disc model in particular, having fixed ζ and α, which determine the exponent of the power law, the parameter ν determines the average degree. In the binomial model, there is an additional dependence on β. Our results focus on the binomial model and show that clustering *does not depend* on ν. Therefore, in the binomial model the "amount" of clustering and the average degree can be tuned *independently*.

1.2   Notation

Let $\{X_N\}_{N \in \mathbb{N}}$ be a sequence of real-valued random variables on a sequence of probability spaces $\{(\Omega_N, \mathbb{P}_N)\}_{N \in \mathbb{N}}$. For a real number a, we write $X_N \overset{p}{\to} a$ or else X_N converges to a *in probability*, if for every $\varepsilon > 0$, we have $\mathbb{P}_N(|X_N - a| > \varepsilon) \to 0$ as $N \to \infty$. If $\mathcal{E}_N$ is a measurable subset of Ω_N, for any $N \in \mathbb{N}$, we say that the sequence $\{\mathcal{E}_N\}_{N \in \mathbb{N}}$ occurs *asymptotically almost surely (a.a.s.)* if $\mathbb{P}(\mathcal{E}_N) = 1 - o(1)$, as $N \to \infty$. However, with a slight abuse of terminology, we will be saying that an *event occurs a.a.s.* implicitly referring to a sequence of events.

For two functions $f, g : \mathbb{N} \to \mathbb{R}$ we write $f(N) \ll g(N)$ if $f(N)/g(N) \to 0$ as $N \to \infty$. Similarly, we will write $f(N) \asymp g(N)$, meaning that there are positive constants c_1, c_2 such that for all $N \in \mathbb{N}$ we have $c_1 g(N) \leq f(N) \leq c_2 g(N)$. Analogously, we write $f(N) \lesssim g(N)$ (resp. $f(N) \gtrsim g(N)$) if there is a positive constant c such that for all $N \in \mathbb{N}$ we have $f(N) \leq c g(N)$ (resp. $f(N) \geq c g(N)$). These are shorthands for the standard Landau notation, but we chose to express them as above in order to make our presentation more readable.

2   Some Geometric Aspects of the Two Models

The disc model on the hyperbolic plane can be also viewed within the framework of random geometric graphs. Within this framework, the disc model may be defined for *any* threshold distance r_N and not merely for threshold distance equal to R. However, only taking $r_N = R$ yields a random graph with constant average degree that is bounded away from 0. More specifically for any $\delta \in (0,1)$, if $r_N = (1-\delta)R$, then the resulting random graph becomes rather trivial and most vertices have no neighbours. On the other hand, if $r_N = (1+\delta)R$, the resulting random graph becomes too dense and its average degree grows polynomially fast in N.

The proof of these observations relies on the following lemma which provides a characterization of what it means for two points u, v to have $d(u,v) \le (1+\delta)R$, for $\delta \in (-1,1)$, in terms of their *relative angle*, which we denote by $\theta_{u,v}$. For this lemma, we need the notion of the *type* of a vertex. For a vertex $v \in V_N$, if r_v is the distance of v from the origin, that is, the radius of v, then we set $t_v = R - r_v$ – we call this quantity the *type* of vertex v. It is not very hard to see that the type of a vertex is approximately exponentially distributed. If we substitute $R - t$ for r in (1), then assuming that t is fixed that expression becomes asymptotically equal to $\alpha e^{-\alpha t}$.

The lemma that connects the hyperbolic distance between two vertices with the relative angle between them is a generalisation of a similar lemma that appears in [6].

Lemma 1. *Let $\delta \in (-1,1)$ be a real number. For any $\varepsilon > 0$ there exists an $N_0 > 0$ and a $c_0 > 0$ such that for any $N > N_0$ and $u, v \in \mathcal{D}_R$ with $t_u + t_v < (1 - |\delta|)R - c_0$ the following hold.*

- *If $\theta_{u,v} < 2(1-\varepsilon)\exp\left(\frac{\zeta}{2}(t_u + t_v - (1-\delta)R)\right)$, then $d(u,v) < (1+\delta)R$.*
- *If $\theta_{u,v} > 2(1+\varepsilon)\exp\left(\frac{\zeta}{2}(t_u + t_v - (1-\delta)R)\right)$, then $d(u,v) > (1+\delta)R$.*

Let us consider temporarily the (modified) disc model, where we assume that two vertices are joined precisely when their hyperbolic distance is at most $(1+\delta)R$. Let $u \in V_N$ be a vertex and assume that $t_u < C$ (by the above observation on the distribution of the type of a vertex, it is not hard to see that most vertices will satisfy this, if C is chosen large). We will show that if $\delta < 0$, then the expected degree of u, in fact, tends to 0. Let us consider a simple case where $0 < \zeta/\alpha < 2$ and δ satisfies $\frac{\zeta}{2\alpha} < 1 - |\delta| < 1$. It can be shown that a.a.s. there are no vertices of type much larger than $\frac{\zeta}{2\alpha}R$. Hence, since $t_u < C$, if N is sufficiently large, then we have $\frac{\zeta}{2\alpha}R < (1 - |\delta|)R - t_u - c_0$. By Lemma 1, the probability that a vertex v has type at most $\frac{\zeta}{2\alpha}R$ and it is adjacent to u (that is, its hyperbolic distance from u is at most $(1+\delta)R$) is proportional to $e^{\frac{\zeta}{2}(t_u + t_v - (1-\delta)R)}/\pi$. If we average this over t_v we obtain

$$\Pr[u \text{ is adjacent to } v | t_u] \asymp \frac{e^{\zeta t_u/2}}{e^{\frac{\zeta}{2}(1-\delta)R}} \int_0^{\frac{\zeta}{2\alpha}R} e^{\zeta t_v/2} \frac{\alpha \sinh(\alpha(R - t_v))}{\cosh(\alpha R) - 1} dt_v$$

$$\lesssim \frac{e^{\zeta t_u/2}}{e^{\frac{\zeta}{2}(1-\delta)R}} \int_0^R e^{\zeta t_v/2} \frac{e^{\alpha(R - t_v)}}{\cosh(\alpha R) - 1} dt_v$$

$$\asymp \frac{e^{\zeta t_u/2}}{e^{\frac{\zeta}{2}(1-\delta)R}} \int_0^R e^{(\zeta/2 - \alpha)t_v} dt_v \overset{0 < \zeta/\alpha < 2}{\asymp} \frac{e^{\zeta t_u/2}}{N^{1-\delta}} \overset{\delta \le 0}{=} o\left(\frac{1}{N}\right).$$

Hence, the probability that there is such a vertex is $o(1)$. Markov's inequality implies that with high probability most vertices will have no neighbors.

A similar calculation can actually show that the above probability is $\Omega\left(\frac{e^{\zeta t_u/2}}{N^{1-\delta}}\right)$. Thereby, if $0 < \delta < 1$, then the expected degree of u is of order N^δ. A more detailed argument can show that the resulting random graph is too dense in the sense that the number of edges is *no longer proportional* to the number of vertices but grows much faster than that.

3   The Clustering Coefficient

The theme of this work is the study of clustering in $\mathcal{G}(N; \zeta, \alpha, \beta, \nu)$. The notion of clustering was introduced by Watts and Strogatz [30], as a measure of the local density of the graph. In the context of biological or social networks, this measures the likelihood of two vertices that have a common neighbor to be joined with each other. This is expressed by the density of the neighborhood of each vertex. More specifically, for each vertex v of a graph, the *local clustering coefficient* is defined to be the density of the neighborhood of v. In [30], the *clustering coefficient* of a graph G, which we denote by $C_1(G)$, is defined as the average of the local clustering coefficients over all vertices of G. The clustering coefficient $C_1(\mathcal{G}(N; \zeta, \alpha, \beta, \nu))$, as a function of β is discussed in [22], where simulations and heuristic calculations indicate that C_1 can be tuned by β. For the disc model, Gugelmann et al. [19] have shown rigorously that this quantity is asymptotically with high probability bounded away from 0 when $0 < \zeta/\alpha < 2$.

The case where $\beta > 1$ and $0 < \zeta/\alpha < 2$ is of particular interest. More specifically, in this regime $\mathcal{G}(N; \zeta, \alpha, \beta, \nu)$ has constant (i.e., not depending on N) average degree that depends on ν, ζ, α and β, whereas the degree distribution follows the tail of a power law with exponent $2\alpha/\zeta + 1$. This has been shown by the second author in [17]. Note that since $2\alpha/\zeta > 1$, the exponent of the power law may take any value greater than 2. When $1 < \zeta/\alpha < 2$, this exponent is between 2 and 3. In [17] we also show that when $\beta \leq 1$, the average degree of the random graph grows at least logarithmically in N.

As we mentioned above, there has been significant experimental evidence which shows that many networks which arise in applications have degree distributions that follow a power law usually with exponent between 2 and 3 (cf. [1] for example). Also, such networks are typically sparse with only a few nodes of very high degree which are the *hubs* of the network. Thus, in the regime where $\beta > 1$ and $0 < \zeta/\alpha < 2$ the random graph $\mathcal{G}(N; \zeta, \alpha, \beta, \nu)$ appears to exhibit these characteristics. In this work, we explore further the potential of this random graph model as a suitable model for complex networks focusing on the notion of *global clustering* and how this is determined by the parameters of the model.

A first attempt to define this notion was made by Luce and Perry [23], but it was rediscovered more recently by Newman, Strogatz and Watts [27]. Given a graph G, we let $T = T(G)$ be the number of triangles of G and let $\Lambda = \Lambda(G)$ denote the number of *incomplete triangles* of G; this is simply the number of (not necessarily induced) paths having length 2. Then the *global clustering coefficient* $C_2(G)$ of a graph G is defined as

$$C_2(G) := \frac{3T(G)}{\Lambda(G)}. \tag{2}$$

This parameter measures the likelihood that two vertices which share a neighbor are themselves adjacent.

The present work has to do with the value of $C_2(\mathcal{G}(N;\zeta,\alpha,\beta,\nu))$. Our results show exactly how clustering can be tuned by the parameters β, ζ and α only. More precisely, our main result states that this undergoes an abrupt change as β crosses the critical value 1.

Theorem 1. *Let* $0 < \zeta/\alpha < 2$. *If* $\beta > 1$, *then*

$$C_2(\mathcal{G}(N;\zeta,\alpha,\beta,\nu)) \xrightarrow{p} \begin{cases} L_\infty(\beta,\zeta,\alpha), & \text{if } 0 < \zeta/\alpha < 1 \\ 0, & \text{if } 1 \leq \zeta/\alpha < 2 \end{cases},$$

where

$$L_\infty(\beta,\zeta,\alpha) =$$
$$\frac{3}{2} \frac{(\zeta - 2\alpha)^2(\alpha - \zeta)}{(\pi C_\beta)^2} \int_{[0,\infty)^3} e^{\frac{\zeta}{2}(t_u+t_v)+\zeta t_w} g_{t_u,t_v,t_w}(\beta,\zeta) e^{-\alpha(t_u+t_v+t_w)} dt_u dt_v dt_w,$$

with

$$g_{t_u,t_v,t_w}(\beta,\zeta) = \int_{[0,\infty)^2} \frac{1}{z_1^\beta + 1} \frac{1}{z_2^\beta + 1} \frac{1}{\left(e^{\frac{\zeta}{2}(t_w-t_v)}z_1 + e^{\frac{\zeta}{2}(t_w-t_u)}z_2\right)^\beta + 1} dz_1 dz_2$$

and $C_\beta := \frac{2}{\beta \sin(\pi/\beta)}$.
If $\beta \leq 1$, *then*

$$C_2(\mathcal{G}(N;\zeta,\alpha,\beta,\nu)) \xrightarrow{p} 0.$$

The fact that the global clustering coefficient asymptotically vanishes when $\zeta/\alpha \geq 1$ is due to the following: when ζ/α crosses 1 vertices of very high degree appear, which incur an abrupt increase on the number of incomplete triangles with no similar increase on the number of triangles to counterbalance that.

Recall that for a vertex $u \in V_N$, its *type* t_u is defined to be equal to $R - r_u$ where r_u is the radius (i.e., its hyperbolic distance from the origin) of u in $\mathcal{D}_R$. When $1 \leq \zeta/\alpha < 2$, vertices of type larger than $R/2$ appear, which affect the tail of the degree sequence of $\mathcal{G}(N;\zeta,\alpha,\beta,\nu)$. For $\beta > 1$, it was shown in [17] that when $1 \leq \zeta/\alpha < 2$ the degree sequence follows approximately a power law with exponent in $(2,3]$. More precisely, asymptotically as N grows, the degree of a vertex $u \in V_N$ conditional on its type follows a Poisson distribution with parameter equal to $Ke^{\zeta t_u/2}$, where $K = K(\zeta,\alpha,\beta,\nu) > 0$. As we have pointed out in Section 2, when $\zeta/\alpha < 1$, a.a.s. all vertices have type less than $R/2$.

Let us consider, for example, more closely the case $\zeta = \alpha$, where the N points are uniformly distributed on $\mathcal{D}_R$. In this case, the type of a vertex u is approximately exponentially distributed with density $\zeta e^{-\zeta t_u}$. Hence, there are about $Ne^{-\zeta(R/2-\omega(N))} \asymp e^{\zeta\omega(N)}$ vertices of type between $R/2 - \omega(N)$ and $R/2 + \omega(N)$; here $\omega(N)$ is assumed to be a slowly growing function. Now, each of these vertices has degree that is (up to multiplicative constants) at least $e^{\frac{\zeta}{2}(\frac{R}{2}-\omega(N))} = N^{1/2}e^{-\zeta\omega(N)/2}$. Therefore, these vertices' contribution to Λ is at least $e^{\zeta\omega(N)} \times \left(N^{1/2}e^{-\zeta\omega(N)/2}\right)^2 = N$.

Now, if vertex u is of type less than $R/2 - \omega(N)$, its contribution to Λ in expectation is proportional to

$$\int_0^{R/2-\omega(N)} \left(e^{\zeta t_u/2}\right)^2 e^{-\zeta t_u} dt_u \asymp R. \tag{3}$$

As most vertices are indeed of type less than $R/2 - \omega(N)$, it follows that these vertices contribute RN on average to Λ.

However, the amount of triangles these vertices contribute is asymptotically much smaller. Recall that for any two vertices u, v the probability that these are adjacent is bounded away from 0 when $d(u, v) < R$. By Lemma 1 having $d(u, v) < R$ can be expressed saying that the relative angle between u and v is $\theta_{u,v} \lesssim e^{\frac{\zeta}{2}(t_u+t_v-R)}$. Consider three vertices w, u, v which, without loss of generality, satisfy $t_v < t_u < t_w < R/2 - \omega(N)$. Since the relative angle between u and v is uniformly distributed in $[0, \pi]$, it turns out that the probability that u is adjacent to w is proportional to $e^{\frac{\zeta}{2}(t_w+t_u-R)}$; similarly, the probability that v is adjacent to w is proportional to $e^{\frac{\zeta}{2}(t_w+t_v-R)}$. Note that these events are independent. Now, conditional on these events, the relative angle between u and w is approximately uniformly distributed in an interval of length $e^{\frac{\zeta}{2}(t_w+t_u-R)}$. Similarly, the relative angle between v and w is approximately uniformly distributed in an interval of length $e^{\frac{\zeta}{2}(t_w+t_v-R)}$. Hence, the (conditional) probability that u is adjacent to v is bounded by a quantity that is proportional to $e^{\frac{\zeta}{2}(t_u+t_v)}/e^{\frac{\zeta}{2}(t_v+t_w)} = e^{\frac{\zeta}{2}(t_u-t_w)}$. This implies that the probability that u, v and w form a triangle is proportional to $e^{\frac{\zeta}{2}t_w+\zeta t_u+\frac{\zeta}{2}t_v}/N^2$. Averaging over the types of these vertices we have

$$\frac{1}{N^2} \int_0^{R/2-\omega(N)} \int_0^{t_w} \int_0^{t_u} e^{\frac{\zeta}{2}t_w+\zeta t_u+\frac{\zeta}{2}t_v-\zeta(t_v+t_u+t_w)} dt_v dt_u dt_w \asymp \frac{1}{N^2}.$$

Hence the expected number of triangles that have all their vertices of type at most $R/2-\omega(N)$ is only proportional to N. Note that if we take $\alpha > \zeta$, then the above expression is still proportional to N, whereas (3) becomes asymptotically constant giving contribution to Λ that is also proportional to N. This makes the clustering coefficient be bounded away from 0 when $\zeta/\alpha < 1$. Our analysis makes the above heuristics rigorous.

It turns out that the situation is somewhat different if we do not take into consideration high-degree vertices (or, equivalently, vertices that have large type). For any fixed $t > 0$, we will consider the global clustering coefficient of the subgraph of $\mathcal{G}(N; \zeta, \alpha, \beta, \nu)$ that is induced by those vertices that have type at most t. We will denote this by $\widehat{C_2}(t)$. We will show that when $\beta > 1$ then for all $0 < \zeta/\alpha < 2$, the quantity $\widehat{C}_2(t)$ remains bounded away from 0 with high probability. Moreover, we determine its dependence on ζ, α, β.

Theorem 2. *Let $0 < \zeta/\alpha < 2$ and let $t > 0$ be fixed. If $\beta > 1$, then*

$$\widehat{C}_2(t) \xrightarrow{p} L(t; \beta, \zeta, \alpha), \tag{4}$$

where

$$L(t; \beta, \zeta, \alpha) := \frac{6 \int_{[0,t)^3} e^{\frac{\zeta}{2}(t_u + t_v) + \zeta t_w} g_{t_u, t_v, t_w}(\beta, \zeta) e^{-\alpha(t_u + t_v + t_w)} dt_u dt_v dt_w}{(\pi C_\beta)^2 \int_{[0,t)^3} e^{\frac{\zeta}{2}(t_u + t_v) + \zeta t_w} e^{-\alpha(t_u + t_v + t_w)} dt_u dt_v dt_w},$$

where $g_{t_u, t_v, t_w}(\beta, \zeta)$ and C_β are as in Theorem 1.

The most involved part of the proofs, which may be of independent interest, has to do with counting triangles in $\mathcal{G}(N; \zeta, \alpha, \beta, \nu)$, that is, with estimating $T(\mathcal{G}(N; \zeta, \alpha, \beta, \nu))$. In fact, most of our effort is devoted to the calculation of the probability that three vertices form a triangle. Thereafter, a second moment argument, together with the fact that the degree of high-type vertices is concentrated around its expected value, implies that $T(\mathcal{G}(N; \zeta, \alpha, \beta, \nu))$ is close to its expected value.

4 Conclusions

In this contribution, we study the presence of clustering as a result of the hyperbolic geometry of a complex network. We consider the model of Krioukov et al. [22], where the resulting random graph is sparse and its degree distribution follows a power law. We quantify the existence of clustering and, furthermore, for the part of the random network that consists of typical vertices, we show that the clustering coefficient is bounded away from 0. More importantly, we find how does this quantity depend on the parameters of the random graph and show that this can be determined independently of the average degree.

References

1. Albert, R., Barabási, A.-L.: Statistical mechanics of complex networks. Reviews of Modern Physics **74**, 47–97 (2002)
2. Balogh, J., Bollobás, B., Krivelevich, M., Müller, T., Walters, M.: Hamilton cycles in random geometric graphs. Ann. Appl. Probab. **21**(3), 1053–1072 (2011)
3. Barabási, A.-L., Albert, R.: Emergence of scaling in random networks. Science **286**, 509–512 (1999)
4. Barrat, A., Weigt, M.: On the properties of small-world network models. European Physical Journal B **13**(3), 547–560 (2000)
5. Bloznelis, M.: Degree and clustering coefficient in sparse random intersection graphs. Ann. Appl. Probab. **23**(3), 1254–1289 (2013)
6. Bode, M., Fountoulakis, N., Müller, T.: On the component structure of random hyperbolic graphs. in preparation

7. Bollobás, B.: Random graphs. Cambridge University Press, xviii+498 pages (2001)
8. Bollobás, B., Riordan, O.: Mathematical results on scale-free random graphs. In: Bornholdt, S., Schuster, H.G. (eds). Handbook of Graphs and Networks: From the Genome to the Internet, pp. 1-34. Wiley-VCH, Berlin (2003)
9. Bollobás, B., Janson, S., Riordan, O.: Sparse random graphs with clustering. Random Structures Algorithms **38**, 269–323 (2011)
10. Börner, K., Maru, J.T., Goldstone, R.L.: Colloquium Paper: Mapping Knowledge Domains: The simultaneous evolution of author and paper networks. Proc. Natl. Acad. Sci. USA **101**, 5266–5273 (2004)
11. Chung, F., Lu, L.: The average distances in random graphs with given expected degrees. Proc. Natl. Acad. Sci. USA **99**, 15879–15882 (2002)
12. Chung, F., Lu, L.: Connected components in random graphs with given expected degree sequences. Annals of Combinatorics **6**, 125–145 (2002)
13. Chung, F., Lu, L.: Complex Graphs and Networks. AMS, viii+264 pages (2006)
14. Coupechoux, E., Lelarge, M.: How clustering affects epidemics in random networks. In: Proceedings of the 5th International Conference on Network Games, Control and Optimization (NetGCooP 2011), Paris, France, pp. 1–7 (2011)
15. Dorogovtsev, S.N.: Lectures on Complex Networks. Oxford University Press, xi+134 pages (2010)
16. Eggemann, N., Noble, S.D.: The clustering coefficient of a scale-free random graph. Discrete Applied Mathematics **159**(10), 953–965 (2011)
17. Fountoulakis, N.: On the evolution of random graphs on spaces of negative curvature. http://arxiv.org/abs/1205.2923 (preprint)
18. Gilbert, E.N.: Random plane networks. J. Soc. Indust. Appl. Math. **9**, 533–543 (1961)
19. Gugelmann, L., Panagiotou, K., Peter, U.: Random Hyperbolic Graphs: Degree Sequence and Clustering. In: Czumaj, A., Mehlhorn, K., Pitts, A., Wattenhofer, R. (eds.) ICALP 2012, Part II. LNCS, vol. 7392, pp. 573–585. Springer, Heidelberg (2012)
20. Hafner, R.: The asymptotic distribution of random clumps. Computing **10**, 335–351 (1972)
21. Janson, S., Łuczak, T., Ruciński, A.: Random graphs. Wiley-Interscience, xii+333 pages (2001)
22. Krioukov, D., Papadopoulos, F., Kitsak, M., Vahdat, A., Boguñá, M.: Hyperbolic Geometry of Complex Networks. Phys. Rev. E **82**, 036106 (2010)
23. Luce, R.D., Perry, A.D.: A method of matrix analysis of group structure. Psychometrika **14**(1), 95–116 (1949)
24. McDiarmid, C., Müller, T.: On the chromatic number of random geometric graphs. Combinatorica **31**(4), 423–488 (2011)
25. Menczer, F.: Growing and navigating the small world Web by local content. Proc.Natl. Acad. Sci. USA **99**, 14014–14019 (2002)
26. Müller, T.: Personal communication
27. Newman, M.E., Strogatz, S.H., Watts, D.J.: Random graphs with arbitrary degree distributions and their applications. Phys. Rev. E **64**, 026118 (2001)
28. Park, J., Newman, M.E.J.: Statistical mechanics of networks. Phys. Rev. E **70**, 066117 (2004)
29. Penrose, M.: Random Geometric Graphs. Oxford University Press, xiv+330 pages (2003)
30. Watts, D.J., Strogatz, S.H.: Collective dynamics of "small-world" networks. Nature **393**, 440–442 (1998)

Burning a Graph as a Model of Social Contagion

Anthony Bonato[1]([✉]), Jeannette Janssen[2], and Elham Roshanbin[2]

[1] Department of Mathematics, Ryerson University, Toronto, ON M5B 2K3, Canada
abonato@ryerson.ca
[2] Department of Mathematics and Statistics, Dalhousie University,
Halifax, NS B3H 3J5, Canada

Abstract. We introduce a new graph parameter called the burning number, inspired by contact processes on graphs such as graph bootstrap percolation, and graph searching paradigms such as Firefighter. The burning number measures the speed of the spread of contagion in a graph; the lower the burning number, the faster the contagion spreads. We provide a number of properties of the burning number, including characterizations and bounds. The burning number is computed for several graph classes, and is derived for the graphs generated by the Iterated Local Transitivity model for social networks.

1 Introduction

The spread of social influence is an active topic in social network analysis; see, for example, [3,8,13,14,18,19]. A recent study on the spread of emotional contagion in Facebook [16] has highlighted the fact that the underlying network is an essential factor; in particular, in-person interaction and nonverbal cues are not necessary for the spread of the contagion. Hence, agents in the network spread the contagion to their friends or followers, and the contagion propagates over time. If the goal was to minimize the time it took for the contagion to reach the entire network, then which agents would you target with the contagion, and in which order?

As a simplified, deterministic approach to these questions, we consider a new approach involving a graph process which we call *burning*. Burning is inspired by graph theoretic processes like Firefighting [4,7,10], graph cleaning [1], and graph bootstrap percolation [2]. There are discrete time-steps or rounds. Each node is either *burned* or *unburned*; if a node is burned, then it remains in that state until the end of the process. Every round, we choose a node to burn. Once a node is burned in round t, in round $t + 1$, each of its unburned neighbours becomes burned. In every round, we choose one additional unburned node to burn (if such a node is available). The process ends when all nodes are burned. The *burning number* of a graph G, written by $b(G)$, is the minimum number of rounds needed for the process to end. For example, it is straightforward to see that $b(K_n) = 2$. However, even for a relatively simple graph such as the path P_n on n nodes, computing the burning number is more complex; in fact, $b(P_n) = \lceil n^{1/2} \rceil$ as stated below in Theorem 3 (and proven in [6]).

© Springer International Publishing Switzerland 2014
A. Bonato et al. (Eds.): WAW 2014, LNCS 8882, pp. 13–22, 2014.
DOI: 10.1007/978-3-319-13123-8_2

Burning may be viewed as a simplified model for the spread of social contagion in a social network such as Facebook or Twitter. The lower the value of $b(G)$, the easier it is to spread such contagion in the graph G. Suppose that in the process of burning a graph G, we eventually burned the whole graph G in k steps, and for each i, $1 \leq i \leq k$, we denote the node that we burn in the i-th step by x_i. We call such a node simply a *source of fire*. The sequence $(x_1, x_2, \ldots, x_k)$ is called a *burning sequence* for G. With this notation, the burning number of G is the length of a shortest burning sequence for G; such a burning sequence is referred to as *optimal*. For example, for the path P_4 with nodes v_1, v_2, v_3, v_4, the sequence (v_2, v_4) is an optimal burning sequence (See Figure 1). Note that for a graph G with at least two nodes, we have that $b(G) \geq 2$.

Fig. 1. Burning the path P_4 (the open circles represent burned nodes)

The goal of the current paper is to introduce the burning number and explore its core properties. A characterization of burning number via a decomposition into trees is given in Theorem 1. As proven in [6], computing the burning number of a graph is **NP**-complete, even for planar, disconnected, or bipartite graphs. As such, we provide sharp bounds on the burning number for connected graphs, which are useful in many cases when computing the burning number. See Theorem 2.2 for bounds on the burning number. We compute the burning number on the Iterated Local Transitivity model for social networks (introduced in [5]) and grids; see Theorem 8 and Theorem 9, respectively. In the final section, we summarize our results and present open problems for future work.

2   Properties of the Burning Number

In this section, we collect a number of results on the burning number, ranging from characterizations, bounds, to computing the burning number on certain kinds of graphs. We first need some terminology. If G is a graph and v is a node of G, then the *eccentricity* of v is defined as $\max\{d(v, u) : u \in G\}$. The *radius* of G is the minimum eccentricity over the set of all nodes in G. The *center* of G consists of the nodes in G with minimum eccentricity. Given a positive integer k, the *k-th closed neighborhood* of v is defined to be the set $\{u \in V(G) : d(u, v) \leq k\}$ and is denoted by $N_k[v]$; we denote $N_1[v]$ simply by $N[v]$.

We first make the following observation. Suppose that $(x_1, x_2, \ldots, x_k)$, where $k \geq 3$, is a burning sequence for a given graph G. For $1 \leq i \leq k$, the fire spread from x_i will burn only all the nodes within distance $k - i$ from x_i by the end of the k-th step. On the other hand, every node $v \in V(G)$ must be either a source of fire, or burned from at least one of the sources of fire by the end of the k-th step. In other words, any node of G that is not a source of fire must be an element of

$N_{k-i}[x_i]$, for some $1 \le i \le k$. Therefore, we can see that $(x_1, x_2, \ldots, x_k)$ forms a burning sequence for G if and only if the following set equation holds:

$$N_{k-1}[x_1] \cup N_{k-2}[x_2] \cup \ldots \cup N_0[x_k] = V(G). \tag{1}$$

Here is another simple observation. For each pair i and j, with $1 \le i < j \le k$, $d(x_i, x_j) \ge j - i$. Since, otherwise, if $d(x_i, x_j) = l < j - i$, then x_j will be burned at stage $l + i$ ($< j$) which is a contradiction. Hence, we have the following corollary.

Corollary 1. *Suppose that $(x_1, x_2, \ldots, x_k)$ is a burning sequence for a graph G. If for some node $x \in V(G) \setminus \{x_1, \ldots, x_k\}$ and $1 \le j \le k-1$, we have that $N[x] \subseteq N[x_j]$, and for every $i \ne j$, $d(x, x_i) \ge |i - j|$, then $(x_1, \ldots, x_{j-1}, x, x_{j+1}, \ldots, x_k)$ is also a burning sequence for G.*

2.1   Characterizations of Burning Number via Trees

The following theorem provides an alternative characterization of the burning number. Note that through the rest of this paper we consider the burning problem for connected graphs. The *depth* of a node in a rooted tree is the number of edges in a shortest path from the node to the tree's root. The *height* of a rooted tree T is the greatest depth in T. A *rooted tree partition* of G is a collection of rooted trees which are subgraphs of G, with the property that the node sets of the trees partition $V(G)$.

Theorem 1. *Burning a graph G in k steps is equivalent to finding a rooted tree partition into k trees $T_1, T_2, \ldots, T_k$, with heights at most $(k - 1), (k - 2), \ldots, 0$, respectively such that for every $1 \le i, j \le k$ the distance between the roots of T_i and T_j is at least $|i - j|$.*

Proof. Assume that $(x_1, x_2, \ldots, x_k)$ is a burning sequence for G. For all $1 \le i \le k$, after x_i is burned, in each round $t > i$ those unburned nodes of G in the $(t-i)$-neighborhood of x_i will burn. Hence, any node v is burned by receiving fire via a shortest path of burned nodes from a fire source like x_i (this path can be of length zero in the case that $v = x_i$). Hence, we may define a surjective function $f : V(G) \to \{x_1, x_2, \ldots, x_k\}$, with $f(v) = x_i$ if v receives fire from x_i, where i is chosen with the smallest index. Now $\{f^{-1}(x_1), f^{-1}(x_2), \ldots, f^{-1}(x_k)\}$ forms a partition of $V(G)$ such that $G[f^{-1}(x_i)]$ (the subgraph induced by $f^{-1}(x_i)$) forms a connected subgraph of G. Since every node v in $f^{-1}(x_i)$ receives the fire spread from x_i through a shortest path between x_i and v, by deleting extra edges in $G[f^{-1}(x_i)]$ we can make a rooted subtree of G, called T_i with root x_i. Since every node is burned after k steps, the distance between each node on T_i and x_i is at most $k - i$. Therefore, the height of T_i is at most $k - i$.

Conversely, suppose that we have a decomposition of the nodes of G into k rooted subtrees $T_1, T_2, \ldots, T_k$, such that for each $1 \le i \le k$, T_i is of height at most $k - i$. Assume that $x_1, x_2, \ldots, x_k$ are the roots of $T_1, T_2, \ldots, T_k$, respectively, and for each pair i and j, with $1 \le i < j \le k$, $d(x_i, x_j) \ge j - i$. Then $(x_1, x_2, \ldots, x_k)$ is a burning sequence for G, since the distance between any node in T_i and x_i is at most $k - i$. Thus, after k steps the graph G will be burned. $\qquad \square$

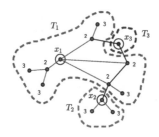

Fig. 2. A rooted tree partition

Figure 2 illustrates Theorem 1. The burning sequence is (x_1, x_2, x_3). We have shown the decomposition of G into subgraphs T_1, T_2, and T_3 based on this burning sequence by drawing dashed curves around the corresponding subgraphs. Each node has been indexed by a number corresponding to the step that it is burned.

The following corollary is useful for determining the burning number of a graph, as it reduces the problem of burning a graph to burning its spanning trees. First, note that for a spanning subgraph H of G, it is evident that $b(G) \le b(H)$ (since every burning sequence for H is also a burning sequence for G).

Corollary 2. *For a graph G we have that*

$$b(G) = \min\{b(T) : T \text{ is a spanning subtree of } G\}.$$

Proof. By Theorem 1, we assume that $T_1, T_2, \ldots, T_k$ is a rooted tree partition of G, where $k = b(G)$, derived from an optimal burning sequence for G. If we take T to be a spanning subtree of G obtained by adding edges sequentially between the T_i's which do not induce a cycle in G, then $b(T) \le k = b(G) \le b(T)$, where the second inequality holds since T is a spanning subgraph of G. $\square$

2.2 Bounds

A subgraph H of a graph G is called an *isometric subgraph* if for every pair of nodes u, v in H, we have that $d_H(u, v) = d_G(u, v)$. For example, a subtree of a tree is an isometric subgraph. As another example, if G is a connected graph and P is a shortest path connecting two nodes of G, then P is an isometric subgraph of G. The following theorem (with proof omitted) shows that the burning number is monotonic on isometric subgraphs.

Theorem 2. *For any isometric subgraph H of a graph G, we have that $b(H) \le b(G)$.*

However, this inequality may fail for non-isometric subgraphs. For example, let H be a path of order 5, and form G by adding a universal node to H. Then $b(H) = 3$, but $b(G) = 2$. The following corollary is an immediate consequence of Theorem 2.

Corollary 3. *If T is a tree and H is a subtree of T, then we have that $b(H) \leq b(T)$.*

The burning number of paths is derived in the following result (with proof omitted).

Theorem 3. *For a path P_n on n nodes, we have that $b(P_n) = \lceil n^{1/2} \rceil$.*

We have the following immediate corollaries.

Corollary 4.   *1. For a cycle C_n, we have that $b(C_n) = \lceil n^{1/2} \rceil$.*
2. For a graph G of order n with a Hamiltonian (that is, spanning) path, we have that $b(G) \leq \lceil n^{1/2} \rceil$.

The following theorem gives sharp bounds on the burning number. For $s \geq 3$, let $K_{1,s}$ denotes a *star*; that is, a complete bipartite graph with parts of order 1 and s. We call a graph obtained by a sequence of subdivisions starting from $K_{1,s}$ a *spider graph*. In a spider graph G, any path which connects a leaf to the node with maximum degree is called an *arm* of G. If all the arms of a spider graph with maximum degree s are of the same length r, we denote such a spider graph by $SP(s, r)$.

Lemma 1. *For any graph G with radius r and diameter d, we have that*

$$\lceil (d + 1)^{1/2} \rceil \leq b(G) \leq r + 1.$$

Proof. Assume that c is a central node of G with eccentricity r. Since every node in G is within distance r from c, the fire will spread to all nodes after $r + 1$ steps. Hence, $r + 1$ is an upper bound for $b(G)$.

Now, let P be a path connecting two nodes u and v in G with $d(u, v) = d$. Since P is an isometric subgraph of G, and $|P| = d + 1$, by Theorem 2 and Theorem 3 we conclude that $b(G) \geq b(P) = \lceil (d + 1)^{1/2} \rceil$.   □

As proven in [6], the lower bound is achieved by paths, and the right side bound is achieved by spider graphs $SP(r, r)$. Note that when proving $b(G) \leq r + 1$ in Theorem 1, we viewed G as covered by a ball with radius r, with a central node chosen as a center of the ball. Hence, by burning a central node, after $r + 1$ steps every node in G will be burned. A *covering* of G is a set of subsets of the nodes of G whose union is $V(G)$. We may generalize this idea to the case that there is a covering of G by a collection of balls with a specified radius.

Theorem 4. *Let $\{C_1, C_2, \ldots, C_t\}$ be a covering of the nodes of a graph G, in which each C_i is a connected subgraph of radius at most k. Then we have that $b(G) \leq t + k$.*

We finish this section by providing some bounds on the burning number in terms of certain domination numbers. A *k-distance dominating set* like D_k for G is a subset of nodes such that for every node $u \in V(G) \setminus D_k$, there exists a node $v \in D_k$, with $d(u, v) \leq k$. The number of the nodes in a minimum k-distance dominating set of G is denoted by $\gamma_k(G)$ and we call it the *k-distance domination number* of G. We have the following result (proof omitted).

Theorem 5. *For any graph G with burning number k we have, $\gamma_{k-1}(G) \leq k$.*

We now give bounds on the burning number in terms of distance domination numbers.

Theorem 6. *If G is a connected graph, then we have that*

$$\frac{1}{2}\Big(\min_{i\geq 1}\{\gamma_i(G) + i\} + 1\Big) \leq b(G) \leq \min_{i\geq 1}\{\gamma_i(G) + i\}.$$

Proof. The upper bound is an immediate corollary of Theorem 4. For the lower bound, let $k = b(G)$, and let $(x_1, \ldots, x_k)$ be a burning sequence. Then we have that

$$V(G) \subseteq N_{k-1}[x_1] \cup \ldots \cup N_0[x_k]$$
$$\subseteq N_{k-1}[x_1] \cup \ldots \cup N_{k-1}[x_k].$$

Hence, $\{x_1, \ldots, x_k\}$ is a k-distance dominating set of G. Since by Theorem 5 we have that $\gamma_{k-1}(G) \leq k$, and $\gamma_{k-1}(G) + (k-1) \leq 2k - 1 = 2b(G) - 1$, we derive that $\min_{i\geq 1}\{\gamma_i(G) + i\} \leq \gamma_{k-1}(G) + (k-1) \leq 2b(G) - 1$.   □

We have the following fact about the k-distance domination number of graphs.

Theorem 7. *[17] If G is a connected graph of order n with $n \geq k+1$, then we have that*

$$\gamma_k(G) \leq \frac{n}{k+1}.$$

Now we use the bound in Theorem 7 for k-distance domination number which provides another upper bound for the burning number.

Corollary 5. *If G is a connected graph of order n, then we have that*

$$b(G) \leq 2n^{1/2} - 1.$$

We conjecture that for any connected graph G of order n, $b(G) \leq \lceil n^{1/2} \rceil$.

3   Burning in the ILT Model

The *Iterated Local Transitivity* (ILT) model [5], simulates on-line social networks (or OSNs). The central idea behind the ILT model is what sociologists call *transitivity*: if u is a friend of v, and v is a friend of w, then u is a friend of w. In its simplest form, transitivity gives rise to the notion of *cloning*, where u is joined to all of the neighbours of v. In the ILT model, given some initial graph as a starting point, nodes are repeatedly added over time which clone *each* node, so that the new nodes form an independent set. The only parameter of the model is the initial graph G_0, which is any fixed finite connected graph. Assume that for a fixed $t \geq 0$, the graph G_t has been constructed. To form G_{t+1}, for each node

$x \in V(G_t)$, add its *clone* x', such that x' is joined to x and all of its neighbours at time t. Note that the set of new nodes at time $t+1$ form an independent set of cardinality $|V(G_t)|$.

The ILT model shares many properties with OSNs such as low average distance, high clustering coefficient densification, and bad spectral expansion; see [5]. The ILT model has also been studied from the viewpoint of competitive diffusion which is one model of the spread of influence; see [20].

We have the following theorem about the burning number of graphs obtained based on ILT model. Even though the graphs generated by the ILT model grow exponentially in order with t, we see that the burning number of such networks remains constant.

Theorem 8. *Let G_t be the graph generated at time $t \geq 1$ based on the ILT model with initial graph G_0. If G_0 has an optimal burning sequence $(x_1, \ldots, x_k)$ in which x_k has a neighbor that is burned in the $(k-1)$-th step, then $b(G_t) = b(G_0)$. Otherwise, $b(G_t) = b(G_0) + 1$.*

Proof. It is straightforward to see that G_0 is an isometric subgraph of G_t. Therefore, by Theorem 2, $b(G_t) \geq b(G_0)$. On the other hand, assume that $(x_1, \ldots, x_k)$ is an optimal burning sequence for G_0. Since every node $x' \in V(G_t) \setminus V(G_0)$ is adjacent to a node in G_0, we have that $(x_1, \ldots, x_k)$ is a burning sequence for the subgraph of G_t induced by $V(G_t) \setminus (N_{G_t}[x_k] \setminus N_{G_0}[x_k])$. Thus, $b(G_t) \leq b(G_0) + 1$. Hence, we conclude that always either we have that $b(G_t) = b(G_0)$, or $b(G_t) = b(G_0) + 1$.

Suppose that for every optimal burning sequence of G_0 all the neighbours of x_k are burned in the k-th step. We claim that $b(G_1) = b(G_0) + 1$. Assume not; that is, $b(G_1) = b(G_0)$. Let $(y_1, y_2, \ldots, y_k)$ be an optimal burning sequence for G_1. Without loss of generality, by Corollary 1, and the structure of G_1, we can assume that $\{y_1, y_2, \ldots, y_{k-1}\} \subseteq G_0$. Then, we have two possibilities; either $y_k = x$ or $y_k = x' \in V(G_1) \setminus V(G_0)$, for some $x \in V(G_0)$. If the former holds, then to burn x' by the end of the k-th step, one of the nodes in the neighbourhood of x must be burned in an earlier stage, which is a contradiction. Since in this case $(y_1, y_2, \ldots, y_k)$ forms a burning sequence for G_0. If the latter holds, that is, $y_k = x' \in V(G_1) \setminus V(G_0)$, for some $x \in V(G_0)$, then, we must have $x = y_{k-1}$ (Note that all the neighbours of x must be burned either in the $(k-1)$-th step or the k-th step; Otherwise, y_k is burned before the k-th step, which is a contradiction). Otherwise, if $x \neq y_{k-1}$, to burn x by the k-th step, one of the neighbours of x must be burned in an earlier stage. But then in this case, $(y_1, \ldots, y_{k-1}, x)$ forms an optimal burning sequence for G_0 such that one of the neighbours of x is burned in the $(k-1)$-th step which is a contradiction with the assumption. Thus, $x = y_{k-1}$.

If all the neighbours of x, including y, are burned in the $(k-1)$-th step, then $(y_1, \ldots, y_{k-2}, y, x)$ forms an optimal burning sequence for G_0. But this is a contradiction with the assumption. If at least one of the neighbours of x like y is burned at the k-th step, then $(y_1, \ldots, y_{k-2}, x, y)$ forms an optimal burning sequence for G_0, which is again a contradiction with the assumption. Therefore, in this case, $b(G_1) = b(G_0)$ is impossible, and hence, $b(G_1) = b(G_0) + 1$.

Conversely, suppose that $b(G_1) = b(G_0) + 1$, and $(x_1, \ldots, x_k)$ is an optimal burning sequence for G_0. If x_k has a neighbour that is burned at stage $k - 1$, then x'_k is also burned at stage k. Therefore, $(x_1, \ldots, x_k)$ is a burning sequence for G_1, and we have that $b(G_1) = b(G_0)$, which is a contradiction. Thus, $b(G_1) = b(G_0)+1$, if and only if for every optimal burning sequence of G_0, say $(x_1, \ldots, x_k)$, all the neighbours of x_k are burned in stage k. By induction, we can conclude that $b(G_t) = b(G_0) + 1$ if and only if for every optimal burning sequence of G_0, say $(x_1, \ldots, x_k)$, all the neighbours of x_k are burned in stage k. Since starting from any graph G_0, for any $t \geq 1$, $b(G_t) = b(G_0)$, or $b(G_t) = b(G_0) + 1$, we conclude that $b(G_t) = b(G_0)$ if and only if for every optimal burning sequence of G_0, say $(x_1, \ldots, x_k)$ one of the neighbours of x_k is burned in stage $k - 1$.   □

4   Cartesian Grids

The *Cartesian product* of graphs G and H, written $G \Box H$, has nodes $V(G) \times V(H)$ with (u, v) adjacent to (x, y) if $u = x$ and $vy \in E(H)$ or $v = y$ and $ux \in E(G)$. The *Cartesian $m \times n$ grid* is $P_m \Box P_n$. We prove the following theorem.

Theorem 9. *If G is a Cartesian $m \times n$ grid with $1 \leq m \leq n$, then we have that*

$$b(G) = \begin{cases} \Theta(n^{1/2}) & \text{if } m = O(n^{1/2}) \\ \Theta((mn)^{1/3}) & \text{if } m = \Omega(n^{1/2}). \end{cases}$$

Proof. First, we find a general upper bound by applying the covering idea in Theorem 4 as follows. Using a layout as shown in Figure 3 we may provide a covering of G by a collection of t closed neighbourhoods of radius r. Note that the r-th neighbourhood of a vertex in a grid is a subset of a "diamond" with diameter $2r+1$ in the Cartesian grid plane. Thus, by a simple counting argument we have that

$$t = \left\lceil \frac{m}{2r+1} \right\rceil \left\lceil \frac{n}{2r+1} \right\rceil + \left(\left\lceil \frac{m}{2r+1} \right\rceil + 1 \right) \left(\left\lceil \frac{n}{2r+1} \right\rceil + 1 \right)$$
$$\leq 2 \left(\left\lceil \frac{m}{2r+1} \right\rceil + 1 \right) \left(\left\lceil \frac{n}{2r+1} \right\rceil + 1 \right).$$

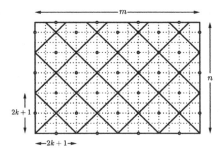

Fig. 3. A covering of the Cartesian grid

Therefore, $t = O(\frac{mn}{r^2} + \frac{m}{r} + \frac{n}{r})$, and consequently, by Theorem 4,

$$b(G) = O\left(r + \frac{mn}{r^2} + \frac{m}{r} + \frac{n}{r}\right). \qquad (2)$$

First, we consider the case that $m = O(n^{1/2})$: Since P_n is an isometric subgraph of G, then by Theorem 3, we have that $b(G) = \Omega(n^{1/2})$. Moreover, by taking $r = n^{1/2}$, we derive that $\frac{mn}{r^2} = m = O(n^{1/2})$, and $\frac{m}{r} + \frac{n}{r} \leq 2\frac{n}{r} = O(n^{1/2})$. Thus, by equation (2), $b(G) = O(n^{1/2})$, and we conclude that in this case, $b(G) = \Theta(n^{1/2})$.

Now, suppose $m = \Omega(n^{1/2})$. Let $S = (x_1, x_2, \ldots, x_k)$ be a burning sequence for G. Thus, every node in G must be in the $(k-i)$-th neighborhood of a node x_i, for some $1 \leq i \leq k$. By direct checking, the number of nodes in the r-th closed neighborhood of a node x in G equals

$$|N_r[x]| = |\{y \in G : d(x, y) \leq r\}| = 1 + 4 + \cdots + 4r$$
$$= 1 + 2r(r+1).$$

Therefore, by double counting the nodes of G and by (1), we have that

$$mn = |G| \leq |N_{k-1}[x_1]| + |N_{k-2}[x_2]| + \cdots + |N_0[x_k]|$$
$$= k + \sum_{i=1}^{k-1} 2i(i+1) = \frac{2k^3 + k}{3}.$$

Since the above inequality holds for all burning sequences, we conclude that $b(G) = \Omega((mn)^{1/3})$. On the other hand, by taking $r = (mn)^{1/3}$ in equation (2), we derive that $b(G) = O((mn)^{1/3})$. Hence, the proof follows.   □

5   Conclusions and Future Work

We introduced a new graph parameter, the burning number of a graph, written $b(G)$. The burning number measures how rapidly social contagion spreads in a given graph. We gave a characterization of the burning number in terms of decompositions into trees, and gave bounds on the burning number which allow us to compute it for a variety of graphs. We determined the asymptotic order of the burning number of grids, and determined the burning number in the Iterated Local Transitive model for social networks.

Several problems remain on the burning number. We conjecture that for a connected graph G of order n, $b(G) \leq \lceil n^{1/2} \rceil$. Determining the burning number remains open for many classes of graphs, including trees and disconnected graphs. It remains open to consider the burning number in real-world social networks such as Facebook or LinkedIn. As Theorem 8 suggests, the burning number of on-line social networks is likely of constant order as the network grows over time. We remark that burning number generalizes naturally to directed graphs; one interesting direction is to determine the burning number on Kleinberg's small world model [15], which adds random directed edges to the Cartesian grid.

A simple variation which leads to complex dynamics is to change the rules for nodes to burn. As in graph bootstrap percolation [2], the rules could be varied so nodes burn only if they are adjacent to at least r burned neighbors, where $r > 1$. We plan on studying this variation in future work.

References

1. Alon, N., Prałat, P., Wormald, N.: Cleaning regular graphs with brushes. SIAM Journal on Discrete Mathematics **23**, 233–250 (2008)
2. Balogh, J., Bollobás, B., Morris, R.: Graph bootstrap percolation (preprint 2014)
3. Banerjee, S., Das, A., Gopalan, A., Shakkottai, S.: Epidemic spreading with external agents. In: Proceedings of IEEE Infocom (2011)
4. Barghi, A., Winkler, P.: Firefighting on a random geometric graph. Random Structures and Algorithms (accepted)
5. Bonato, A., Hadi, N., Horn, P., Prałat, P., Wang, C.: Models of on-line social networks. Internet Mathematics **6**, 285–313 (2011)
6. Bonato, A., Janssen, J., Roshanbin, E.: Burning a graph is hard (preprint 2014)U
7. Bonato, A., Nowakowski, R.J.: The Game of Cops and Robbers on Graphs. American Mathematical Society, Providence (2011)
8. Domingos, P., Richardson, M.: Mining the network value of customers. In: Proceedings of the 7th International Conference on Knowledge Discovery and Data Mining (KDD) (2001)
9. Finbow, S., King, A., MacGillivray, G., Rizzi, R.: The firefighter problem for graphs of maximum degree three. Discrete Mathematics **307**, 2094–2105 (2007)
10. Finbow, S., MacGillivray, G.: The Firefighter problem: a survey of results, directions and questions. Australasian Journal of Combinatorics **43**, 57–77 (2009)
11. Garey, M.R., Johnson, D.S.: Computers and Intractability: A Guide to the Theory of NP-Completeness. Freeman, W.H (1979)
12. Haynes, T.W., Hedetniemi, S.T., Slater, P.J.: Fundamentals of Domination in Graphs. Marcel Dekker, New York (1998)
13. Kempe, D., Kleinberg, J., Tardos, E.: Maximizing the spread of influence through a social network. In: Proceedings of the 9th International Conference on Knowledge scovery and Data Mining (KDD) (2003)
14. Kempe, David, Kleinberg, Jon M., Tardos, Éva: Influential Nodes in a Diffusion Model for Social Networks. In: Caires, Luís, Italiano, Giuseppe F., Monteiro, Luís, Palamidessi, Catuscia, Yung, Moti (eds.) ICALP 2005. LNCS, vol. 3580, pp. 1127–1138. Springer, Heidelberg (2005)
15. Kleinberg, J.: The small-world phenomenon: an algorithmic perspective. In: Proc. 32nd ACM Symp. Theory of Computing (2000)
16. Kramer, A.D.I., Guillory, J.E., Hancock, J.T.: Experimental evidence of massive-scale emotional contagion through social networks. Proceedings of the National Academy of Sciences **111**, 8788–8790 (2014)
17. Meir, A., Moon, J.W.: Relations between packing and covering numbers of a tree. Pacific Journal of Mathematics **61**, 225–233 (1975)
18. Mossel, E., Roch, S.: On the submodularity of influence in social networks. In: Proceedings of 39th Annual ACM Symposium on Theory of Computing (STOC) (2007)
19. Richardson, M., Domingos, P.: Mining knowledge-sharing sites for viral marketing. In: Proceedings of the 8th International Conference on Knowledge scovery and Data Mining (KDD) (2002)
20. Small, L., Mason, O.: Information diffusion on the iterated local transitivity model of online social networks. Discrete Applied Mathematics **161**, 1338–1344 (2013)
21. West, D.B.: Introduction to Graph Theory, 2nd edn. Prentice Hall (2001)

Personalized PageRank
with Node-Dependent Restart

Konstantin Avrachenkov[1]($\boxtimes$), Remco van der Hofstad[2], and Marina Sokol[1]

[1] Inria Sophia Antipolis, Valbonne, France
{k.avrachenkov,msokol}@inria.fr
[2] Eindhoven University of Technology, Eindhoven, The Netherlands
r.w.v.d.hofstad@tue.nl

Abstract. Personalized PageRank is an algorithm to classify the importance of web pages on a user-dependent basis. We introduce two generalizations of Personalized PageRank with node-dependent restart. The first generalization is based on the proportion of visits to nodes before the restart, whereas the second generalization is based on the proportion of time a node is visited just before the restart. In the original case of constant restart probability, the two measures coincide. We discuss interesting particular cases of restart probabilities and restart distributions. We show that both generalizations of Personalized PageRank have an elegant expression connecting the so-called direct and reverse Personalized PageRanks that yield a symmetry property of these Personalized PageRanks.

1 Introduction and Definitions

PageRank has become a standard algorithm to classify the importance of nodes in a network. Let us start by introducing some notation. Let $G = (V, E)$ be a finite graph, where V is the node set and $E \subseteq V \times V$ the collection of (directed) edges. Then, PageRank can be interpreted as the stationary distribution of a random walk on G that restarts from a uniform location in V at each time with fixed probability $1 - \alpha \in (0, 1)$. Thus, in the Standard PageRank centrality measure [10], the random walk restarts after a geometrically distributed number of steps, and the restart takes place from a uniform location in the graph, and otherwise jumps to any one of the neighbours in the graph with equal probability. Personalized PageRank [17] is a modification of the Standard PageRank where the restart distribution is not uniform. Both the Standard and Personalized PageRank have many applications in data mining and machine learning (see e.g., [3,4,10,13,16,17,19,20]).

In the (standard) Personalized PageRank, the random walker restarts with a given fixed probability $1 - \alpha$ at every step, independently of the node the walker presently is at. We suggest a generalization where a random walker restarts with probability $1 - \alpha_i$ when it is at node $i \in V$. When the random walker restarts, it chooses a node to restart at with probability distribution v^T. In many cases, we let the random walker restart at a fixed location, say $j \in V$.

© Springer International Publishing Switzerland 2014
A. Bonato et al. (Eds.): WAW 2014, LNCS 8882, pp. 23–33, 2014.
DOI: 10.1007/978-3-319-13123-8_3

Then the Personalized PageRank is a vector whose ith coordinate measures the importance of node i to node j.

The above random walks $(X_t)_{t \geq 0}$ can be described by a finite-state Markov chain with the transition matrix

$$\tilde{P} = AD^{-1}W + (I - A)\underline{1}v^T, \tag{1}$$

where W is the (possibly non-symmetric) adjacency matrix, D is the diagonal matrix with diagonal entries $d_i = D_{ii} = \sum_{j=1}^{n} W_{ij}$, and $A = \text{diag}(\alpha_1, \dots, \alpha_n)$ is the diagonal matrix of damping factors. The case of undirected graphs corresponds to the case when W is a symmetric matrix. In general, D_{ii} is the out-degree of node $i \in V$. If some node does not have outgoing edges, we add artificial outgoing edges from that node to all the other nodes. Throughout the paper, we assume that the graph is strongly connected, that is, each node can be reached from any other node.

We propose two generalizations of the Personalized PageRank with node-dependent restart:

Definition 1 (Occupation-Time Personalized PageRank, OT PPR).
The Occupation-Time Personalized PageRank *with restart vector v is the vector whose ith coordinate is given by*

$$\pi_i(v) = \lim_{t \to \infty} \mathbb{P}(X_t = i). \tag{2}$$

By the fact that $(\pi_i(v))_{i \in V}$ is the stationary distribution of the Markov chain, we can interpret $\pi_i(v)$ as a long-run frequency of visits to node i, i.e.,

$$\pi_i(v) = \lim_{t \to \infty} \frac{1}{t} \sum_{s=1}^{t} \mathbb{1}_{\{X_s = i\}}. \tag{3}$$

Our second generalization is based on the location where the random walker restarts.

Definition 2 (Location-of-Restart Personalized PageRank, LOR PPR).
The Location-of-Restart Personalized PageRank *with restart vector v is the vector whose ith coordinate is given by*

$$\rho_i(v) = \lim_{t \to \infty} \mathbb{P}(X_t = i \text{ just before restart}) = \lim_{t \to \infty} \mathbb{P}(X_t = i \mid \text{restart at time } t+1). \tag{4}$$

We can interpret $\rho_i(v)$ as a long-run frequency of visits to node i which are immediately followed by a restart, i.e.,

$$\rho_i(v) = \lim_{t \to \infty} \frac{1}{N_t} \sum_{s=1}^{t} \mathbb{1}_{\{X_t = i, X_{t+1} \text{ restarts}\}}, \tag{5}$$

where N_t denotes the number of restarts up to time t. When the restarts occur with equal probability at every node, $N_t \sim \mathsf{Bin}(t, 1 - \alpha)$, i.e., N_t has a binomial distribution with t trials and success probability $1 - \alpha$. When the restart probabilities are unequal, the distribution of N_t is more involved. We note that

$$N_t/t \xrightarrow{a.s.} \sum_{i \in V} (1 - \alpha_i)\pi_i(v), \tag{6}$$

where $\xrightarrow{a.s.}$ denotes convergence almost surely.

Both generalized Personalized PageRanks are probability distributions, i.e., their sum over $i \in V$ gives 1. When $v^T = e(j)$, where $e_i(j) = 1$ when $i = j$ and $e_i(j) = 0$ when $i \neq j$, then both $\pi_i(v)$ and $\rho_i(v)$ can be interpreted as the relative importance of node i from the perspective of node j.

We see at least three applications of the generalized Personalized PageRank. The network sampling process introduced in [6] can be viewed as a particular case of PageRank with a node-dependent restart. We discuss this relation in more detail in Section 4. Secondly, the generalized Personalized PageRank can be applied as a proximity measure between nodes in semi-supervised machine learning [5,16]. In this case, one may prefer to discount the effect of less informative nodes, e.g., nodes with very large degrees. And thirdly, the generalized Personalized PageRank can be applied for spam detection and control. It is known [11] that spam web pages are often designed to be ranked highly. By using the Location-of-Restart Personalized PageRank and penalizing the ranking of spam pages with small restart probability, one can push the spam pages from the top list produced by search engines.

Let us mention some other works generalizing the fixed probability of restart in PageRank. In [14] and [15] the authors consider the damping factor as a random variable distributed according to user behavior. These works generalize [9] where the random damping factor is chosen according to the uniform distribution. Also, there is a stream of works, starting from [8], that generalize the damping parameter to the damping function. In those works, the random walk restarts with probability as a function of the number of steps from the last restart.

In this paper, we investigate the two generalizations of Personalized PageRank. The paper is organised as follows. In Section 2, we investigate the Occupation-Time Personalized PageRank. In Section 3, we investigate the Location-of-Restart Personalized PageRank. In Section 4, we specify the results for some particular interesting cases. We close in Section 5 with a discussion of our results and suggestions for future research.

All proofs can be found in the accompanying research report [1].

2   Occupation-Time Personalized PageRank

The Occupation-Time Personalized PageRank can be calculated explicitly as follows:

Theorem 1 (Occupation-Time Personalized PageRank Formula). *The Occupation-Time Personalized PageRank $\pi(v)$ with restart vector v and node-dependent restart equals*

$$\pi(v) = \frac{1}{v^T[I - AP]^{-1}\underline{1}}v^T[I - AP]^{-1}, \tag{7}$$

with $P = D^{-1}W$ the transition matrix of random walk on G without restarts.

By the renewal-reward theorem (see e.g., Theorem 2.2.1 in [22]), formula (7) admits the following probabilistic interpretation

$$\pi_i(v) = \frac{\mathbb{E}_v[\# \text{ visits to } i \text{ before restart}]}{\mathbb{E}_v[\# \text{ steps before restart}]}, \tag{8}$$

where $\mathbb{E}_v$ denotes expectation with respect to the Markov chain starting in distribution v.

Denote for brevity $\pi_i(j) = \pi_i(e_j^T)$, where e_j is the jth vector of the standard basis, so that $\pi_i(j)$ denotes the importance of node i from the perspective of node j. Similarly, $\pi_j(i)$ denotes the importance of node j from the perspective of i. We next provide a relation between these "direct" and "reverse" PageRanks in the case of *undirected* graphs.

Theorem 2 (Symmetry for undirected Occupation-Time Personalized PageRank). *When $W^T = W$ and $A > 0$, the following relation holds*

$$\frac{d_j}{\alpha_j K_j(A)}\pi_i(j) = \frac{d_i}{\alpha_i K_i(A)}\pi_j(i), \tag{9}$$

with $d_i = D_{ii}$ the degree of node i and

$$K_i(A) = \frac{1}{e_i^T[I - AP]^{-1}\underline{1}}. \tag{10}$$

We note that the term $(AD^{-1}W)^k$ can be interpreted as the contribution corresponding to all paths of length k, while $K_i(A)$ can be interpreted as the reciprocal of the expected time between two consecutive restarts if the restart distribution is concentrated on node i, i.e.,

$$K_i(A)^{-1} = \mathbb{E}_i[\# \text{ steps before restart}], \tag{11}$$

see also (8). Thus, a probabilistic interpretation of (9) is that

$$\frac{d_j}{\alpha_j}\mathbb{E}_j[\# \text{ visits to } i \text{ before restart}] = \frac{d_i}{\alpha_i}\mathbb{E}_i[\# \text{ visits to } j \text{ before restart}]. \tag{12}$$

Since

$$\mathbb{E}_i[\# \text{ visits to } j \text{ before restart}] = \sum_{k=1}^{\infty} \sum_{v_1,\dots,v_k} \prod_{t=0}^{k-1} \frac{\alpha_{v_s}}{d_{v_s}}, \tag{13}$$

where $v_0 = i$, we immediately see that the expression for

$$\mathbb{E}_j[\# \text{ visits to } i \text{ before restart}]$$

is identical to (13), except for the first factor of α_i/d_i, which is present in

$$\mathbb{E}_i[\# \text{ visits to } j \text{ before restart}],$$

but not in $\mathbb{E}_j[\# \text{ visits to } i \text{ before restart}]$, and the factor α_j/d_j, which is present in $\mathbb{E}_j[\# \text{ visits to } i \text{ before restart}]$, but not in $\mathbb{E}_i[\# \text{ visits to } j \text{ before restart}]$. This explains the factors d_i/α_i and d_j/α_j in (12) and gives a probabilistic proof alternative to the algebraic proof given in [1].

3   Location-of-Restart Personalized PageRank

The Location-of-Restart Personalized PageRank can also be calculated explicitly:

Theorem 3 (Location-of-Restart Personalized PageRank Formula).
The Location-of-Restart Personalized PageRank $\rho(v)$ with restart vector v and node-dependent restart is equal to

$$\rho(v) = v^T[I - AP]^{-1}[I - A], \qquad (14)$$

with $P = D^{-1}W$.

The proof follows from (13) and the formula

$$\rho_i(v) = \mathbb{E}_v[\# \text{ visits to } i \text{ before restart}]\mathbb{P}(\text{restart from } i) \qquad (15)$$
$$= \mathbb{E}_v[\# \text{ visits to } i \text{ before restart}](1 - \alpha_i).$$

The Location-of-Restart Personalized PageRank admits an even more elegant relation between the "direct" and "reverse" PageRanks in the case of undirected graphs:

Theorem 4 (Symmetry for undirected Location-of-Restart Personalized PageRank).   *When $W^T = W$ and $\alpha_i \in (0,1)$, the following relation holds*

$$\frac{1 - \alpha_j}{\alpha_j} \, d_j \, \rho_i(j) = \frac{1 - \alpha_i}{\alpha_i} \, d_i \, \rho_j(i). \qquad (16)$$

A probabilistic proof follows from (12) and (15).

Interestingly, in (9), the whole graph topology has an effect on the relation between the "direct" and "reverse" Personalized PageRanks, whereas in the case of $\rho(v)$, see equation (16), only the local end-point information (i.e., α_i and d_i) have an effect on the relation between the "direct" and "reverse" PageRanks. We have no intuitive explanation of this distinction.

4   Interesting Particular Cases

In this section, we consider some interesting particular cases for the choice of restart probabilities and distributions.

4.1   Constant Probability of Restart

The case of constant restart probabilities (i.e., $\alpha_i = \alpha$ for every i) corresponds to the original or standard Personalized PageRank. We note that in this case the two generalizations coincide. For instance, we can recover a known formula [21] for the original Personalized PageRank with $A = \alpha I$ from equation (7). Specifically,

$$v^T[I - AP]^{-1}\underline{1} = v^T[I - \alpha P]^{-1}\underline{1} = v^T \sum_{k=0}^{\infty} \alpha^k P^k \underline{1} = \frac{1}{1 - \alpha}, \tag{17}$$

and hence we retrieve the well-known formula

$$\pi(v) = (1 - \alpha)v^T[I - \alpha P]^{-1}. \tag{18}$$

We also retrieve the following elegant result connecting "direct" and "reverse" original Personalized PageRanks on undirected graphs ($W^T = W$) obtained in [5]:

$$d_i \pi_j(i) = d_j \pi_i(j), \tag{19}$$

since in the original Personalized PageRank $\alpha_i \equiv \alpha$. Finally, we note that in the original Personalized PageRank, the expected time between restarts does not depend on the graph structure nor on the restart distribution and is given by

$$\mathbb{E}_v[\text{time between consecutive restarts}] = \frac{1}{1 - \alpha}, \tag{20}$$

which is just the mean of a geometrically distributed random variable with parameter $1 - \alpha$.

4.2   Restart Probabilities Proportional to Powers of Degrees

Let us consider a particular case when the restart probabilities are proportional to powers of the degrees. Namely, let $\sigma \in \mathbb{R}$ and define

$$A = I - aD^{\sigma}, \tag{21}$$

with $ad_{\max}^{\sigma} < 1$. We first analyse $[I - AP]^{-1}$ with the help of a Laurent series expansion. Let $T(\varepsilon) = T_0 - \varepsilon T_1$ be a substochastic matrix for small values of ε and let T_0 be a stochastic matrix with associated stationary distribution ξ^T and deviation matrix $H = (I - T_0 + \underline{1}\xi^T)^{-1} - \underline{1}\xi^T$. Then, the following Laurent series expansion takes place (see Lemma 6.8 from [2])

$$[I - T(\varepsilon)]^{-1} = \frac{1}{\varepsilon}X_{-1} + X_0 + \varepsilon X_1 + \dots, \tag{22}$$

where the first two coefficients are given by

$$X_{-1} = \frac{1}{\pi^T T_1 \underline{1}} \underline{1} \xi^T,$$ (23)

and

$$X_0 = (I - X_{-1}T_1)H(I - T_1 X_{-1}).$$ (24)

Applying the above Laurent power series to $[I - AP]^{-1}$ with $T_0 = P$, $T_1 = D^\sigma P$ and $\varepsilon = a$, we obtain

$$[I - AP]^{-1} = [I - (P - aD^\sigma P)]^{-1} = \frac{1}{a}\frac{1}{\pi^T T_1 \underline{1}}\underline{1}\xi^T + O(1)$$ (25)

$$= \frac{1}{a}\frac{1}{\xi^T D^\sigma \underline{1}}\underline{1}\xi^T + O(1).$$

This yields the following asymptotic expressions for the generalized Personalized PageRanks

$$\pi_j = \xi_j + O(a),$$ (26)

and

$$\rho_j = \frac{d_j^\sigma \xi_j}{\sum_{i \in V} d_i^\sigma \xi_i} + O(a).$$ (27)

In particular, if we assume that the graph is undirected ($W^T = W$), then $\xi_j = d_j/\sum_i d_i$ and we can further specify the above expressions as

$$\pi_j = \frac{d_j}{\sum_i d_i} + O(a),$$ (28)

and

$$\rho_j = \frac{d_j^{1+\sigma}}{\sum_{i \in V} d_i^{1+\sigma}} + O(a).$$ (29)

We observe that using a positive or negative power σ of the degrees, we can significantly penalize or promote the score ρ for nodes with large degrees.

As a by-product of our computations, using (11), we have also obtained a nice asymptotic expression for the expected time between restarts in the case of undirected graph:

$$\mathbb{E}_v[\text{time between consecutive restarts}] = \frac{1}{a}\frac{\sum_{i \in V} d_i}{\sum_{i \in V} d_i^{1+\sigma}} + O(1).$$ (30)

One interesting conclusion from the above expression is that when $\sigma > 0$ the highly skewed distribution of the degree in G can significantly shorten the time between restarts.

4.3   Random Walk with Jumps

In [6], the authors introduced a process with artificial jumps. It is suggested in [6] to add artificial edges with weights a/n between each two nodes to the graph. This process creates self-loops as well. Thus, the new modified graph is a combination of the original graph and a complete graph with self-loops. Let us demonstrate that this is a particular case of the introduced generalized definition of Personalized PageRank. Specifically, we define the damping factors as

$$\alpha_i = \frac{d_i}{d_i + a}, \quad i \in V, \tag{31}$$

and as the restart distribution we take the uniform distribution $(v = \underline{1}/n)$. Indeed, it is easy to check that we retrieve the transition probabilities from [6]

$$p_{ij} = \begin{cases} \frac{a+n}{n(d_i+a)} & \text{when } i \text{ has an edge to } j, \\ \frac{a}{n(d_i+a)} & \text{when } i \text{ does not have an edge to } j. \end{cases} \tag{32}$$

As was shown in [6], the stationary distribution of the modified process, coinciding with the Occupation-Time Personalized PageRank, is given by

$$\pi_i = \pi_i(\underline{1}/n) = \frac{d_i + a}{2|E| + na}, \quad i \in V. \tag{33}$$

In particular, from (6) we conclude that in the stationary regime

$$\mathbb{E}_\pi[\text{time between consecutive restarts}] = \left(\sum_{j \in V} \left(1 - \frac{d_j}{d_j + a} \right) \frac{d_j + a}{2|E| + na} \right)^{-1}$$

$$= \frac{2|E| + na}{na} = \frac{\bar{d} + a}{a},$$

where $\bar{d}$ is the average degree of the graph. Since $\pi(v)$ is the stationary distribution of $\tilde{P}$ with $v = \underline{1}/n$ (see (1)), it satisfies the equation

$$\pi(AP + [I - A]\underline{1}v^T) = \pi. \tag{34}$$

Rewriting this equation as

$$\pi[I - A]\underline{1}v^T = \pi[I - AP], \tag{35}$$

and postmultiplying by $[I - AP]^{-1}$, we obtain

$$\pi[I - A]\underline{1}v^T[I - AP]^{-1} = \pi \tag{36}$$

or

$$v^T[I - AP]^{-1} = \frac{\pi}{\sum_{j=1}^{n} \pi_j(1 - \alpha_j)}. \tag{37}$$

This yields

$$\rho_i(v) = \frac{\pi_i(1 - \alpha_i)}{\sum_{j=1}^n \pi_j(1 - \alpha_j)}. \tag{38}$$

In our particular case of $\alpha_i = d_i/(d_i + a)$, the combination of (33) and (38) gives that $\pi_i(1 - \alpha_i)$ is independent of i, so that

$$\rho_i = 1/n. \tag{39}$$

This is quite surprising. Since $v^T = \frac{1}{n}\mathbf{1}^T$, the nodes just after restart are distributed uniformly. However, it appears that the nodes just before restart are also uniformly distributed! Such effect has also been observed in [7]. Algorithmically, this means that all pages receive the *same* generalized Personalized PageRank ρ, which, for ranking purposes, is rather uninformative. On the other hand, this Personalized PageRank can be useful for sampling procedures. In fact, we can generalize (31) to

$$\alpha_i = \frac{d_i}{d_i + a_i}, \quad i \in V, \tag{40}$$

where now each node has its own parameter a_i. Now it is convenient to take as the restart distribution

$$v_i = \frac{a_i}{\sum_{k \in V} a_k}.$$

Performing similar calculations as above, we arrive at

$$\pi_i(v) = \frac{d_i + a_i}{2|E| + \sum_{k \in V} a_k}, \quad i \in V,$$

and

$$\rho_i(v) = \frac{a_i}{\sum_{k \in V} a_k}, \quad i \in V.$$

Now in contrast with (39), the Location-of-Restart Personalized PageRank can be tuned to give any distribution that we like.

5   Discussion

We have proposed two generalizations of Personalized PageRank when the probability of restart depends on the node. Both generalizations coincide with the original Personalized PageRank when the probability of restart is the same for all nodes. However, in general they show quite different behavior. In particular, the Location-of-Restart Personalized Pagerank appears to be stronger affected by the value of the restart probabilities. We have further suggested several applications of the generalized Personalized PageRank in machine learning, sampling and information retrieval and analyzed some particularly interesting cases.

We feel that the analysis of the generalized Personalized PageRank on random graph models is a promising future research direction. We have already obtained some indications that the degree distribution can strongly affect the time between restarts. It would be highly interesting to analyze this effect in more detail on various random graph models (see e.g., [18] for an introduction into random graphs, and [12] for first results on directed configuration models).

Acknowledgments. The work of KA and MS was partially supported by the EU project Congas and Alcatel-Lucent Inria Joint Lab. The work of RvdH was supported in part by Netherlands Organisation for Scientific Research (NWO). This work was initiated during the 'Workshop on Modern Random Graphs and Applications' held at Yandex, Moscow, October 24-26, 2013. We thank Yandex, and in particular Andrei Raigorodskii, for bringing KA and RvdH together in such a wonderful setting.

References

1. Avrachenkov, K., van der Hofstad, R.W., Sokol, M.: Personalized PageRank with Node-dependent Restart, Inria Research Report no.8570 (July 2014). http://hal.inria.fr/INRIA-RRRT/hal-01052482
2. Avrachenkov, K., Filar, J., Howlett, P.: Analytic perturbation theory and its applications. SIAM Publisher (2013)
3. Avrachenkov, K., Dobrynin, V., Nemirovsky, D., Pham, S., Smirnova, E.: Pagerank based clustering of hypertext document collections. In: Proceedings of ACM SIGIR 2008 (2008)
4. Avrachenkov, K., Gonçalves, P., Mishenin, A., Sokol, M.: Generalized optimization framework for graph-based semi-supervised learning. In: Proceedings of SIAM Conference on Data Mining (SDM 2012)
5. Avrachenkov, K., Gonçalves, P., Sokol, M.: On the Choice of Kernel and Labelled Data in Semi-supervised Learning Methods. In: Bonato, A., Mitzenmacher, M., Prałat, P. (eds.) WAW 2013. LNCS, vol. 8305, pp. 56–67. Springer, Heidelberg (2013)
6. Avrachenkov, K., Ribeiro, B., Towsley, D.: Improving Random Walk Estimation Accuracy with Uniform Restarts. In: Kumar, R., Sivakumar, D. (eds.) WAW 2010. LNCS, vol. 6516, pp. 98–109. Springer, Heidelberg (2010)
7. Avrachenkov, K., Litvak, N., Sokol, M., Towsley, D.: Quick detection of nodes with large degrees. Internet Mathematics **10**, 1–19 (2013)
8. Baeza-Yates, R., Boldi, P., Castillo, C.: Generalizing pagerank: Damping functions for link-based ranking algorithms. In: Proceedings of ACM SIGIR 2006, pp. 308–315 (2006)
9. Boldi, P.: TotalRank: Ranking without damping. In: Poster Proceedings of WWW 2005, pp. 898–899 (2005)
10. Brin, S., Page, L., Motwami, R., Winograd, T.: The PageRank citation ranking: bringing order to the Web. Stanford University Technical Report (1998)
11. Castillo, C., Donato, D., Gionis, A., Murdock, V., Silvestri, F.: Know your neighbors: Web spam detection using the web topology. In: Proceedings of ACM SIGIR 2007, pp. 423–430 (July 2007)
12. Chen, N., Olvera-Cravioto, M.: Directed random graphs with given degree distributions. Stochastic Systems 3, 147–186 (electronic) (2013)
13. Chen, P., Xie, H., Maslov, S., Redner, S.: Finding scientific gems with Google's PageRank algorithm. Journal of Informetrics **1**(1), 8–15 (2007)
14. Constantine, P.G., Gleich, D.F.: Using Polynomial Chaos to Compute the Influence of Multiple Random Surfers in the PageRank Model. In: Bonato, A., Chung, F.R.K. (eds.) WAW 2007. LNCS, vol. 4863, pp. 82–95. Springer, Heidelberg (2007)

15. Constantine, P.G., Gleich, D.F.: Random alpha PageRank. Internet Mathematics **6**(2), 189–236 (2010)
16. Fouss, F., Francoisse, K., Yen, L., Pirotte, A., Saerens, M.: An experimental investigation of kernels on graphs for collaborative recommendation and semi-supervised classification. Neural Networks **31**, 53–72 (2012)
17. Haveliwala, T.: Topic-Sensitive PageRank. In: Proceedings of WWW 2002 (2002)
18. van der Hofstad, R.: Random Graphs and Complex Networks, Lecture notes in preparation (2014) (preprint). http://www.win.tue.nl/~rhofstad/NotesRGCN.html
19. Liu, X., Bollen, J., Nelson, M.L., van de Sompel, H.: Co-authorship networks in the digital library research community. Information Processing & Management **41**, 1462–1480 (2005)
20. Massa, P., Avesani, P.: Trust-aware recommender systems. In: Proceedings of the 2007 ACM Conference on Recommender Systems (RecSys 2007), pp. 17–24 (2007)
21. Moler, C.D., Moler, K.A.: Numerical Computing with MATLAB. SIAM (2003)
22. Tijms, H.C.: A first course in stochastic models. John Wiley and Sons (2003)

Efficient Computation of the Weighted Clustering Coefficient

Silvio Lattanzi[1] and Stefano Leonardi[2](✉)

[1] Google Research, New York, USA
[2] Sapienza University of Rome, Roma, Italy
leonardi@dis.uniroma1.it

Abstract. The clustering coefficient of an unweighted network has been extensively used to quantify how tightly connected is the neighbor around a node and it has been widely adopted for assessing the quality of nodes in a social network. The computation of the clustering coefficient is challenging since it requires to count the number of triangles in the graph. Several recent works proposed efficient sampling, streaming and MapReduce algorithms that allow to overcome this computational bottleneck. As a matter of fact, the intensity of the interaction between nodes, that is usually represented with weights on the edges of the graph, is also an important measure of the statistical cohesiveness of a network. Recently various notions of weighted clustering coefficient have been proposed but all those techniques are hard to implement on large-scale graphs.

In this work we show how standard sampling techniques can be used to obtain efficient estimators for the most commonly used measures of weighted clustering coefficient. Furthermore we also propose a novel graph-theoretic notion of clustering coefficient in weighted networks.

1 Introduction

In recent years we observed a growing attention on the study of the structural properties of social networks [15,17] as result of the fast increase of the amount of social network data available for research. A widely adopted measure of the graph structure of a social network is the clustering coefficient [30]. The local clustering coefficient of a node is defined as the probability that any two neighbors of a node are themselves neighbors. The clustering coefficient of a graph is the average local clustering coefficient of the nodes of the graph.

The clustering coefficient is used to measure how tightly interconnected the community is around a node. The degree of closeness of any two neighbors of a node is also interpreted as an index of trust of the node itself. The local clustering coefficient of a node has been proved for example to be a relevant feature for detecting spam nodes in the web [3] and high quality users in social networks [3].

Work partially done while visiting scientist at Google Research NY. Partially supported from Google Focused Award "Algorithms for Large-scale Data Analysis", EU FET project MULTIPLEX 317532, EU ERC project PAAI 259515.

© Springer International Publishing Switzerland 2014
A. Bonato et al. (Eds.): WAW 2014, LNCS 8882, pp. 34–46, 2014.
DOI: 10.1007/978-3-319-13123-8_4

Computing the clustering coefficient of a network is a challenging computational task since it reduces to counting the number of triangles in a graph. This task can be naively executed in $O(n^3)$ time or it can be reduced to matrix multiplication. The problem of computing the local clustering coefficient for every node of the network is even more challenging. Several recent works have proposed a variety of efficient methods for fast computation of clustering coefficient in large scale networks based on random sampling [10], streaming algorithms [6,13], and MapReduce parallel computation [27].

However, most of the studies on the structural properties of social networks have focused on unweighted networks. In practice, many real world networks exhibit a varying degree of intensity and heterogeneity in the connections which is usually modeled with positive real weights on edges. Weights on edges are used for instance to measure the number of messages exchanged between friends or the number of links between hosts. Since the statistical level of cohesiveness in a network should in principle depend also on the weight of the edges, some recent interesting papers started to investigate weighted networks [19]. Several new notions of weighted clustering coefficient have also been introduced ([2, 21] among others) but, unfortunately, no efficient method for estimating the weighted clustering coefficient has been presented so far.

Computing the exact values of the weighted clustering coefficient is at least as hard as for the unweighted clustering coefficient. Sampling is the key for an efficient and accurate approximation [6,10].

Our Contributions. We summarize in the following the main contributions of our work:

1. We show how to obtain efficient estimators for several standard definitions of weighted clustering coefficient. Our sampling algorithm are easily parallelizable too.
2. We introduce a novel notion of *weighted clustering coefficient*. We base our proposal on the observation that edges with large weights are more likely to play a role in the social network. Our model defines a family of unweighted random graphs with edges existing with different probabilities. The probability of an edge depends on its weight. The largest the weight, the highest the probability. Each graph of the family of random graphs is an unweighted graph. The local *weighted clustering coefficient* of a node is defined as the expected local clustering coefficient in the family of random graphs. Our definition naturally extends to the weighted clustering coefficient of the entire graph[1]. We also design a polynomial time algorithm to compute the value of the *weighted clustering coefficient* and a sampling technique to estimate it efficiently.
3. We perform experiments that show interesting properties of the weighted clustering coefficient.

---

[1] We note that our definition of weighted random random graph is different from the definition of [9] and it is more in line with the standard definition used in data mining and biology [11].

1.1   Related Works

A survey of several approaches to clustering coefficient in weighted networks can be found in [24]. In [22] the definition of clustering coefficient is based on the average weight on the edges of a triangle. In [2] the definition of the local clustering coefficient of a node only depends on the weights of the two edges incident to the node but not on the weight of the third edge of the triangle. In [1] it is adopted the standard unweighted definition with the exception that triangles are weighted by the edge that closes the triangle. In [21] the weight is only considered in the numerator of the definition of clustering coefficient whereas the denominator is the one of the unweighted case. In [31] the weight of a triangle is obtained by multiplying the weights of the edges. Other proposals that are substantially different from our approach can also be found in [14,32]. The study of the clustering coefficient in several classes of random unweighted graphs can be found in [4].

The problem of estimating the clustering coefficient is closely related to the problem of counting the number of triangles in a graph. This is computationally expensive even on graphs of moderate size because of the time complexity needed to enumerate all the length-two paths of the graph. Several works proposed efficient heuristics [16,26] with computational results reported for graphs of large size. More recently, there are algorithms designed under the MapReduce [8] programming model. Using a MapReduce infrastructure, [27] proposed algorithms for computing the exact number of triangles and the clustering coefficient of graphs. Randomized algorithms for counting triangles were also implemented under the MapReduce paradigm [23]. Finally to estimate the total number of triangle in a graph is possible to use also matrix sketches [18], unfortunately it is not clear how to extend this approach to local clustering coefficient. A related measure is also the transitivity coefficient of a graph [20]. Techniques adopted for estimating the clustering coefficient usually extend to the transitivity coefficient.

A natural approach for problems in massive networks is also to provide approximate solutions based on the application of data stream and random sampling algorithms. These algorithms usually provide an $(1 \pm \epsilon)$ approximation of the number of triangles with probability $1 - \delta$. The number of samples and amount of memory needed depends on the quality of the approximation. Data stream algorithms for estimating the number of triangles of a graph have been considered in [13,29]. Semi-streaming algorithms have been proposed in [3]. A sampling-based algorithm for estimating the clustering coefficient of a graph is given in [25].

2   Preliminaries

Let $G = (V, E)$ be an undirected graph with $n = |V|$ and $m = |E|$ edges. For every vertex $v \in V$ let $\mathcal{N}(v, G)$ denote its neighborhood, i.e. $\mathcal{N}(v, G) = \{u \in V : \exists (u, v) \in E\}$. The *clustering coefficient* $C_v(G)$ of a vertex $v \in V$ is defined as the probability that a random pair of its neighbors is connected by an edge, i.e. $C_v(G) := \frac{\left|\{(u,w)\in E:u,w\in\mathcal{N}(v,G)\}\right|}{\binom{|\mathcal{N}(v,G)|}{2}}$.   In case of $|\mathcal{N}(v,G)| < 2$ we define $C_v(G) := 0$. The *clustering coefficient* $C(G)$ of G is the average clustering coefficient of its vertices, i.e. $C(G) = \frac{1}{n} \cdot \sum_{v\in V} C_v(G)$.

Let us denote by $W(v,G) = \{\langle u,w\rangle : u,w \in \mathcal{N}(v,G)\}$ the set of *wedges* of vertex v in graph G, i.e., the set of distinct paths of length two centered at v.

We denote by $w : E \to \Re^+$ the positive weight on the edges of the graph. Let $W = \max_{e\in E} w(e)$ be the maximum weight of an edge. We normalize the edge weights in a way that their range varies in $[0,1]$. Denote by $p : E \to [0,1]$ the normalized weights. We denote with $\mathbf{1}_C$, the indicator variable for the event C. In the experimental section we will use the following classic normalization $p(e) = \frac{1}{1+\log W/w(e)}$.

Finally, we say that we have an (ϵ,δ) estimator for a measure M, if we can estimate M within an ϵ multiplicative factor with probability at least $1 - \delta$.

2.1 Generalizations of Clustering Coefficient in Weighted Networks

In this paper we consider three generalizations of the clustering coefficient in weighted networks. In particular we focus our attention to two definitions proposed in [2,21] that well represent two general approaches to the problem: in one case the weights of the edges are added, in the other case they are multiplied. We additionally introduce a novel definition that is particularly relevant when the weights on the edges can be interpreted as probabilities[2].

Onnela et al. The first definition of clustering coefficient that we consider has been introduced by Onnela et al. [21]:

$$WC_v^{Onnela} = \frac{\sum_{\langle u,w\rangle\in W(v,G)} \hat{w}(e(v,u))\hat{w}(e(v,w))\hat{w}(e(u,w))}{|\mathcal{N}(v,G)|\,(|\mathcal{N}(v,G)| - 1)}.$$

where with $w(e(v,u))$ we indicate the weight of the edge $e(v,u)$ and $\hat{w}(e(\cdot,\cdot)) = \frac{w(e(\cdot,\cdot))}{W}$.

Barrat et al. The second definition of clustering coefficient that we consider has been introduced by Barrat et al. [2]:

$$WC_v^{Barrat} = \frac{\sum_{\langle u,w\rangle\in W(v,G)}(w(e(v,u)) + w(e(v,w)))\mathbf{1}_{e(u,w)}}{(|\mathcal{N}(v,G)| - 1)\sum_{v\in e} w(e)}.$$

where $\mathbf{1}_{e(u,w)}$ is equal 1 if the edge (u,w) exists and 0 otherwise.

---

[2] This setting is particularly relevant generated using inference models [12].

Weighted clustering coefficient for probabilistic networks. The last measure that we analyze is novel. The basic idea is that the normalized weights can be interpreted as probabilities of existence of the edges in the graph. More formally, define the class of random graph $\mathcal{G}_{n,p}$ with edge e appearing independently with probability $p(e)$. Each graph $G' = (V, E') \in \mathcal{G}_{n,p}$ is an edge subset E' of E. The probability of G' is $p(G') = \prod_{e \in E'} p(e) \prod_{e \notin E'} (1 - p(e))$.

The *weighted clustering coefficient* WC_v of a vertex $v \in V$ is defined as the expected clustering coefficient over the class of graphs $\mathcal{G}_{n,p}$: $WC_v^{random} = \mathrm{E}_{G' \in \mathcal{G}_{n,p}} C_v(G')$.

3   Computing the Weighted Clustering Coefficient in Probabilistic Networks

In this section we give a polynomial algorithm to compute the new definition of weighted clustering coefficient efficiently. Note that at first sight our problem seems computationally very challenging because there are exponentially many possible realizations of the neighborhood of each node.[3]

Our first algorithmic contribution is to show that the problem is in P, we give an algorithm with complexity $O(|\mathcal{N}(v, G)|^4)$. Our algorithm is based on a dynamic program that computes incrementally the contribution of each neighbor pair to the clustering coefficient of each node.

Unfortunately our exact algorithm is too slow to run on real networks where the maximum degree is typically very large(in the order of millions for Twitter or Google+) fortunately in the next section we show that the new measure has an efficient (ϵ, δ) estimator.

Recall that the unweighted clustering coefficient of a node v is defined as the probability that a randomly selected pair of its neighbors is connected by an edge, based on this we can give an alternative definition of weighted clustering coefficient for probabilistic networks. Let $\chi(u, w)$ be a random variable that has value 1 if the randomly selected pair is (u, w) and 0 otherwise. We have: $C_v(G) := \sum_{u,w \in \mathcal{N}(v,G) \wedge (u,w) \in E} Pr(\chi(u, w) = 1)$. Where each pair is counted only once. In the following we shorten $\mathcal{N}(v, G')$ to $\mathcal{N}'(v)$. Using this definition we can rewrite the weighted clustering coefficient for v as: $WC_v^{random} = \mathrm{E}_{G' \in \mathcal{G}_{n,p}} \left[\sum_{u,w \in \mathcal{N}'(v) \wedge (u,w) \in E'} Pr(\chi(u, w) = 1 | G') \right].$

Now by defining $\xi(u, w)$ a random value that has value 1 if and only if $u, w \in \mathcal{N}'(v) \wedge (u, w) \in E'$, and by denoting with $\mathbf{1}_{\xi(u,w)}$ its indicator function, we have:

---

[3] Note that enumerating all the triangles in the graph would not work in this setting because of the dependency induced by the number of wedges in the realization of the random graph.

$$WC_v^{random}$$

$$= \mathbb{E}_{G' \in \mathcal{G}_{n,p}} \left[\sum_{u,w \in \mathcal{N}'(v) \wedge (u,w) \in E'} Pr(\chi(u,w) = 1 | G') \right]$$

$$= \mathbb{E}_{G' \in \mathcal{G}_{n,p}} \left[\sum_{u,w \in \mathcal{N}(v)} \left(\mathbf{1}_{\xi(u,w)} Pr(\chi(u,w) = 1 | G') \right) \right]$$

$$= \sum_{u,w \in \mathcal{N}(v)} \mathbb{E}_{G' \in \mathcal{G}_{n,p}} \left[\mathbf{1}_{\xi(u,w)} Pr(\chi(u,w) = 1 | G') \right]$$

$$= \sum_{u,w \in \mathcal{N}(v)} \left(Pr(\xi(u,w) = 1) * \mathbb{E}_{G' \in \mathcal{G}_{n,p}} \left[\mathbf{1}_{\xi(u,w)} Pr(\chi(u,w) = 1 | G') \Big| \xi(u,w) = 1 \right] \right)$$

$$= \sum_{u,w \in \mathcal{N}(v)} \left(Pr(u,w \in \mathcal{N}'(v) \wedge (u,w) \in E') * Pr(\chi(u,w) = 1 | \xi(u,w) = 1) \right)$$

Now the first term of the sum can be easily computed because $Pr(u,w \in \mathcal{N}'(v) \wedge (u,w) \in E') = p(e_{u,v})p(e_{w,v})p(e_{w,u})$. The second term is still problematic. In fact $Pr(\chi(u,w) = 1 | \xi(u,w) = 1)$ depends on all the possible instantiations of G' and so it potentially involve the computation of exponentially many terms.

In the following we show how to compute it efficiently using dynamic programming[4]. Note that $Pr(\chi(u,w) = 1 | \xi(u,v) = 1) = Pr(\chi(u,w) = 1 | u,w \in \mathcal{N}'(v))$ because the existence of the edge (u,w) does not change the probability of selecting u and v as random neighbors of v. And $Pr(\chi(u,w) = 1 | u,w \in \mathcal{N}'(v))$ is the probability that a pair u,w of neighbors of v are selected conditioned on the fact that $u,w \in \mathcal{N}'(v)$.

To compute this probability we use the equivalence between the following two processes. The first one selects two elements uniformly at random without replacement from a set S, and the second one computes a random permutation of the elements in the set S and then returns the first two elements of the permutation.

Using this equivalence we can rephrase the probability $Pr(\chi(u,w) = 1 | u,w \in \mathcal{N}'(v))$ as the probability that in a random permutation of the nodes in $\mathcal{N}(v)$, u and w are the two nodes with the smallest positions in $\mathcal{N}'(v)$. Note that for this to happen either u and w are the first two nodes in the permutation of the nodes in $\mathcal{N}(v)$, or all the nodes that are in positions smaller than u and w do not appear in $\mathcal{N}'(v)$.

Now leveraging on this fact, we give a quadratic dynamic program to compute $Pr(\chi(u,w) = 1 | \xi(u,w) = 1)$. Consider an arbitrary order to the nodes in $\mathcal{N}(v) \setminus \{u,w\}$, $z_1, z_2, ..., z_{|\mathcal{N}(v)|-2}$. In our algorithm we implicitly construct all

---

[4] Unfortunately to the best of our knowledge, there is no analytic technique to estimate this quantity correctly or with a close approximation without using a dynamic programming.

the possible permutations incrementally and at the same time we estimate the probability that u, w are selected in each permutation. More specifically, initially we analyze the permutations containing only the elements $\{u, w\}$ then the ones containing the elements $\{u, w, z_1\}$, then the ones containing the elements $\{u, w, z_1, z_2\}$, and so on so for until we get the probability for each permutation containing all the elements in $\mathcal{N}(v)$.

The key ingredient of our algorithm is the following observation. Once we have computed the probability for all the permutations containing the nodes $\{u, w, z_1, z_2, ..., z_{i-1}\}$, to extend our computation to the permutations containing also the node z_i, we have to consider only two scenarios: in the first one z_i appears after u, w in the permutation in this case the probability that u and w are the nodes in $\mathcal{N}'(v)$ with the two smallest positions is the same. In the second one z_i appears before either of u or of w, conditioned on this event the probability that u and w are the nodes in $\mathcal{N}'(v)$ with the two smallest positions decreases by a multiplicative factor $1 - p(e_{v,z_i})$.

We are now ready to state our dynamic program more formally. Let M be a square matrix of dimension $|\mathcal{N}(v)| - 1$ such that $M_{i,j}$, for $j \leq i$, contains the probability that in a random permutation of nodes $\{u, w, z_1, z_2, ..., z_i\}$ u and w are preceded by exactly j elements in the permutation but they are in the first and second position when we consider the ordering induced only to nodes in $N'(v)$. Note that $M_{0,0}$ is equal to 1, because in this case we consider permutations containing only $\{u, w\}$. Similarly, we can compute $M_{1,0}$ and $M_{1,1}$. In particular, for $M_{1,0}$ we require that z_1 is in a position after u and w so we have $M_{1,0} = \frac{1}{3}M_{0,0}$. Instead, $M_{1,1} = \frac{2}{3}(1 - p(e_{v,z_i}))M_{0,0}$. More generally, we have that for $j \leq i$:

$$
M_{i,j} = \begin{cases} \frac{i-1}{i+1}M_{i-1,0} & \text{if j} = 0 \\ \frac{i-j-1}{i+1}M_{i-1,j} + \frac{j+1}{i+1}\overline{p}(e_{v,z_i})M_{i-1,j-1} & \text{if j<i} \\ & \text{and j>0} \\ \frac{i}{i+1}\overline{p}(e_{v,z_i})M_{i-1,j-1} & \text{if j = i} \end{cases}
$$

Where $\overline{p}(*) = 1 - p(*)$.

Once we have computed the matrix M we can compute $Pr(\chi(u, w) = 1 | u, w \in \mathcal{N}'(v))$, in fact we have that: $Pr(\chi(u, w) = 1 | u, w \in \mathcal{N}'(v)) = \sum_{i=0}^{|\mathcal{N}'(v)|-2} M_{|\mathcal{N}'(v)|-2,i}$ So we have:

$$
WC_v^{random} = \sum_{u,w \in \mathcal{N}(v)} \left(\frac{1}{2}p(e_{u,v})p(e_{w,v})p(e_{w,u}) * \left(\sum_{i=0}^{|\mathcal{N}'(v)|-2} M_{|\mathcal{N}'(v)|-2,i} \right) \right)
$$

Note that the above algorithm has complexity $O(|\mathcal{N}(v)|^4)$, so it is too slow to run on large networks for this reason in the next section we study of efficient estimators.

4   Efficient Estimators for the Weighted Clustering Coefficient

We propose efficient (ϵ, δ) estimators for the various definition of weighted clustering coefficient. Our estimators that use basic concentration theory are similar to the one presented in [5,28]. They are the first linear estimators for the weighted clustering coefficient to the best of our knowledge.

Onnela et al. Recall the definition of Onnela et al. [21] given in Section 2.

In this definition the weighted clustering coefficient is equal to the total normalized weight of the triangles containing v averaged by the number of wedges centered on v. Thus if we sample the wedges uniformly at random, using the Hoeffding bound and the fact that the normalized weights are in $[0, 1]$, we get an efficient (ϵ, δ) estimator for WC_v^{Onnela}[5].

More formally, let $X_1, \ldots, X_s$ identical random variables that with probability $\frac{1}{|\mathcal{N}(v,G)|(|\mathcal{N}(v,G)|-1)}$ have value $\hat{w}(e(v,u))\,\hat{w}(e(v,w))\hat{w}(e(u,w))$ for every wedge $< u, w >$. Then, $E\left[\sum_{i=1}^{s} X_i\right] = sWC_v^{Onnela}$. Furthermore, by Hoeffding bound we have that: $P\left(\left|X - E\left[\sum_{i=1}^{s} X_i\right]\right| \leq \epsilon E\left[\sum_{i=1}^{s} X_i\right]\right) \leq e^{\frac{\epsilon^2 E\left[\sum_{i=1}^{s} X_i\right]}{3}}$

$= e^{\frac{\epsilon s WC_v^{Onnela}}{3}}$ So if we want $\delta > e^{\frac{\epsilon s WC_v^{Onnela}}{3}}$, it suffices to have $s \in O(\log \frac{1}{\delta} \cdot \frac{1}{\epsilon^2 \cdot WC_v^{Onnela}})$ samples.

Lemma 1. *There is a sampling-based algorithm which with probability $1 - \delta$ returns a $(1 \pm \epsilon)$-approximation of the local weighted clustering coefficient WC_v^{Onnela} of a vertex v of a weighted graph G. It needs $\mathcal{O}(\log \frac{1}{\delta} \cdot \frac{1}{\epsilon^2 \cdot WC_v^{Onnela}})$ samples.*

Furthermore note that for the sampler we only need to be able to sample random wedges and this can be easily done in linear time.

Barrat et al. Recall the definition of Barrat et al. [2] given in Section 2. In this case the weighted clustering coefficient is not an explicit average so we cannot use the Hoeffding bound directly as before. Nevertheless note that we can define WC_v^{Barrat} as the average value of the random variable X where X has value $\mathbf{1}_{e(u,w)}$ with probability $\frac{w(e(v,u))+w(e(v,w))}{(|\mathcal{N}(v,G)|-1)\sum_{v\in e} w(e)}$ for all $\langle u, w \rangle \in W(v, G)$.

Using this alternative definition combined with the Chernoff bound we get that by using k samples of the wedges weighted with the correct probability we can get good estimation of WC_v^{Barrat}. We omit the formal proof of the correctness of this estimator for lack of space and because it is very similar to the previous one.

---

[5] Note that for this to work it is fundamental that the weight on the edges have been normalized and so are in $[0, 1]$.

Lemma 2. *There is a sampling-based algorithm which with probability $1 - \delta$ returns a $(1 \pm \epsilon)$-approximation of the local weighted clustering coefficient WC_v^{Barrat} of a vertex v of a weighted graph G. It needs $\mathcal{O}(\log \frac{1}{\delta} \cdot \frac{1}{\epsilon^2 \cdot WC_v^{Barrat}})$ samples.*

Weighted clustering coefficient for probabilistic networks. The algorithm is based on sampling a random wedge $\langle u, w \rangle \in W(v, G')$ from a random graph $G' \in \mathcal{G}_{n,p}$ and checking whether $(u, w) \in G'$. The core idea of our sampler is to generate for a node v s neighbor realizations $\mathcal{N}(v)_1, \ldots, \mathcal{N}(v)_s$ uniformly at random from $\mathcal{G}_{n,p}$. Then for each realization sample a random wedge $< u, w >$ uniformly from $\mathcal{N}(v)_i$ and check if the wedge is part of a triangle in the realization. The estimation of the clustering coefficient is equal to the number of wedge that are part of a triangle divided by s.

For the sake of completeness we give a simple analysis of the algorithm below. We first show that the expected value of X_i is exactly WC_v^{random}.

We have for each $i \in \{1, \ldots, s\}$: $\mathrm{E}\left[X_i\right] = \mathrm{E}_{G' \in \mathcal{G}_{n,p}} \left[\frac{\left| \left\{ (u,w) \in E' : u, w \in \mathcal{N}(v, G') \right\} \right|}{\binom{|\mathcal{N}(v,G')|}{2}} \right]$

$= \mathrm{E}_{G' \in \mathcal{G}_{n,p}} [C_v(G')] = WC_v^{random}$

Then we use the fact that for $0 - 1$ random variables we have $\mathbf{Var}\left[X_i\right] \leq \mathbf{E}[X_i^2] = \mathbf{E}[X_i] = WC_v^{random}$.

Now we analyze the variance of X. Since the X_i are mutually independent we get $\mathbf{Var}\left[X\right] = \mathbf{Var}\left[\frac{1}{s} \cdot \sum_{i=1}^{s} X_i\right] = \frac{1}{s^2} \cdot \sum_{i=1}^{s} \mathbf{Var}\left[X_i\right] \leq \frac{WC_v^{random}}{s}$. Finally, we can apply Chebyshev inequality. This gives us $\mathbf{Pr}\left[\left|X - \mathbf{E}[X]\right| \geq \epsilon \cdot \mathbf{E}[X]\right] \leq \frac{\mathbf{Var}[X]}{(\epsilon \cdot \mathbf{E}[X])^2} \leq \frac{WC_v^{random}}{s \cdot \epsilon^2 \cdot (WC_v^{random})^2} = \frac{1}{s \cdot \epsilon^2 \cdot WC_v^{random}}$.

If $s \geq \frac{3}{\epsilon^2 \cdot WC_v^{random}}$ then with probability $\frac{2}{3}$ the algorithm *sampling WC_v^{random}* approximates the weighted clustering coefficient of vertex v in G within a relative error of $(1 \pm \epsilon)$. In order to amplify the probability of success we run the algorithm $\Theta(\log \frac{1}{\delta})$ times and return the median of all results. This leads to the following corollary:

Lemma 3. *There is a sampling-based algorithm which with probability $1 - \delta$ returns a $(1 \pm \epsilon)$-approximation on the local weighted clustering coefficient WC_v^{random} of a vertex v of a weighted graph G. It needs $\mathcal{O}\left(\log \frac{1}{\delta} \cdot \frac{1}{\epsilon^2 \cdot WC_v^{random}}\right)$ samples.*

We note that all the previous estimation algorithms are implementable using 2 rounds of MapReduce. The first round samples random wedges and to check if they form triangles, and the second round cumulates the scores. Furthermore the total amount of messages in each round is bounded by $\mathcal{O}\left(\sum_v \log \frac{1}{\delta} \cdot \frac{1}{\epsilon^2 \cdot WC_v^*}\right)$, and all the mappers and the reducers have $\tilde{O}(n)$ running time, where n is the size of their input. Additional details of the MapReduce implementation are omitted in the extended abstract for lack of space.

5   Experiments

The main goal of this section is to show experimentally some properties of the weighted clustering coefficient and to show the speed-up obtained by our simple estimators.

Dataset and experiment settings. The clustering coefficient is a fundamental topological property of networks and also one of the most used topological features in machine learning on graphs. Indeed, it has been used to detect spam on the web [3].

For this reason we study the effectiveness of weighted clustering coefficient by studying its power as a machine learning feature. In particular, we focus on the specific case where we are interested in detecting spam in the Web. Toward this end, we use a public available dataset [7] composed by a collection of hosts manually labelled (spam/non spam) by a group of volunteers and by the weighted host graph network. The graph is composed by 114,529 hosts in the .UK domain and there are 5709 hosts marked as "non spam" and 344 hosts marked as "spam". Even if the web graph is directed in this section we ignore the directionality of the edges for simplicity[6]. Finally we note that there are 2058 hosts marked as "non spam" and 93 hosts marked as "spam" with clustering coefficient bigger than 0 (for any, weighted or unweighted, definition of clustering coefficient).

In our experiments we are only interested in analyze the correlation between various definitions of clustering coefficient and the integrity of an host. To do it, for each definition we first compute the corresponding score for each labelled node, then we rank all the labelled nodes with score bigger than 0 according to their scores and compute the precision of each position i of the ranking as the percentage of "non spam" hosts before position i. This measure, even if simplistic, gives a good intuition of the correlation between the clustering coefficient and the goodness of a page.

Performances of the sampling algorithm. Here we first analyze the running time of the sampling algorithm presented in Section 4 when we vary the number of samples used in the algorithm and we compare its running time with the running time of the algorithms that considers all the triangles to compute the unweighted clustering coefficient or the weighted clustering coefficient defined by Barrat et al. [2].

Then we analyze how the precision of the ranking varies as a function of the number of samples performed by the algorithm. In Figure 1 we show the average running time of the sampling algorithm when we vary the number of samples. It is interesting to note that the running time increase almost linearly with the number of seeds showing that the algorithm efficiently use all the parallelization offered by the MapReduce framework. Furthermore it is quite interesting to note the huge difference in running time between the sampling algorithm and

---

[6] Note that all the discussed notion of clustering coefficient can be extended to capture the directionality of the edges.

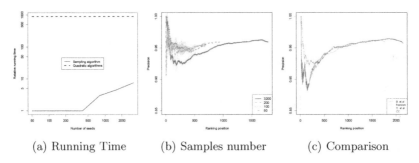

(a) Running Time   (b) Samples number   (c) Comparison

Fig. 1. Running time *vs* Number of seeds and precision *vs* Ranking position of the sample algorithm when we vary the number of samples between 50, 100, 200 and 3200 and comparison with unweighted clustering coefficient

the quadratic algorithm that considers all the triangles. In fact when we used 50, 100, 200 and 400 samples the sampling algorithm is 900 times faster than the quadratic algorithm, and even when we use 3200 samples the sampling algorithm is still 120 times faster!

Now we turn our attention to the effects of varying the number of samples on the precision of the algorithm. In Figure 1 we show how the precision curve of the new notion of weighted clustering coefficient changes when we use 50, 100, 200 or 3200 samples(we notice a similar trends also with 400, 800, 1600 samples and for other clustering coefficient definition, we do not show them in the figure for clarity). Also in this case we plot the average precisions with lines and the standard deviations with the shadows around the lines. From the plots it seems that few samples are enough to obtain a good estimation of the weighted clustering coefficient.

Finally we compare the weighted clustering coefficient with the classic clustering coefficient. It is possible to note that the ranking obtained by our new definition Random of weighted clustering coefficient has higher precision for the first positions in the ranking while it has performances comparable with the rankings obtained using the other definitions on the higher positions of the ranking. The definitions of Onnela et al. and Barrat et al. have performances very similar to the classic unweighted definition of clustering coefficient.

Acknowledgments. We thank Corinna Cortes for suggesting the problem.

References

1. Applying social network analysis to the information in cvs repositories. In: 1st International Workshop on Mining Software Repositories (MSR)
2. Barrat, A., Barthlemy, M., Pastor-Satorras, R., Vespignani, A.: The architecture of complex weighted networks. Proceedings of the National Academy of Sciences of the United States of America

3. Becchetti, L., Boldi, P., Castillo, C., Gionis, A.: Efficient semi-streaming algorithms for local triangle counting in massive graphs. In: KDD 2008 (2008)
4. Bollobs, B.: Mathematical results on scale-free random graphs. In: Handbook of Graphs and Networks
5. Budak, C., Agrawal, D., El Abbadi, A.: Structural trend analysis for online social networks. In: VLDB 2011 (2011)
6. Buriol, L., Frahling, G., Leonardi, S., Marchetti-Spaccamela, A., Sohler, C.: Counting triangles in data streams. In: PODS 2006 (2006)
7. Castillo, C., Donato, D., Becchetti, L., Boldi, P., Leonardi, S., Santini, M., Vigna, S.: A reference collection for web spam. SIGIR 2006 (2006)
8. Dean, J., Ghemawat, S.: Mapreduce: Simplified data processing on large clusters. In: OSDI 2004 (2004)
9. Fagiolo, G.: Clustering in complex directed networks. Phys. Rev. E.
10. Hardiman, S.J., Katzir, L.: Estimating clustering coefficients and size of social networks via random walk. In: WWW 2013 (2013)
11. Hintsanen, P., Toivonen, H.: Finding reliable subgraphs from large probabilistic graphs. Data Min. Knowl. Discov.
12. Hintsanen, P., Toivonen, H.: Finding reliable subgraphs from large probabilistic graphs. Data Min. Knowl. Discov. (2008)
13. Jha, M., Seshadhri, C., Pinar, A.: A space efficient streaming algorithm for triangle counting using the birthday paradox. In: KDD 2013 (2013)
14. Kalna, G., Higham, D.J.: Clustering coefficients for weighted networks. In: Symposium on Network Analysis in Natural Sciences and Engineering
15. Kwak, H., Lee, C., Park, H., Moon, S.: What is twitter, a social network or a news media?. In: WWW 2010 (2010)
16. Latapy, M.: Main-memory triangle computations for very large (sparse(power-law)) graphs. Theoretical Computer Science
17. Leskovec, J., Horvitz, E.: Planetary-scale views on a large instant-messaging network. In: WWW 2008 (2008)
18. Liberty, E.: Simple and deterministic matrix sketches. In: KDD 2014 (2014)
19. Newman, M.E.J.: Analysis of weighted networks. Phys. Rev. E **70**, 056131 (2004)
20. Newman, M.E.J., Watts, D.J., Strogatz, S.H.: Random graph models of social networks. Proc. Natl. Acad. Sci. USA **99**, 2566–2572 (2002)
21. Onnela, J.-P., Saramäki, J., Kertész, J., Kaski, K.: Intensity and coherence of motifs in weighted complex networks. Physical Review E
22. Opsahl, T., Panzarasa, P.: Clustering in weighted networks. Social Networks
23. Pagh, R., Tsourakakis, C.E.: Colorful triangle counting and a mapreduce implementation
24. Saramäki, J., Kivelä, M., Onnela, J.-P., Kaski, K., Kertesz, J.: Generalizations of the clustering coefficient to weighted complex networks. Physical Review E
25. Schank, T., Wagner, D.: Approximating clustering coefficient and transitivity. Journal of Graph Algorithms and Applications
26. Schank, Thomas, Wagner, Dorothea: Finding, Counting and Listing All Triangles in Large Graphs, an Experimental Study. In: Nikoletseas, Sotiris E. (ed.) WEA 2005. LNCS, vol. 3503, pp. 606–609. Springer, Heidelberg (2005)
27. Suri, S., Vassilvitskii, S.: Counting triangles and the curse of the last reducer. In: WWW 2011 (2011)
28. Tsourakakis, C.E., Kang, U., Miller, G.L., Faloutsos, C.: Doulion: counting triangles in massive graphs with a coin. In: KDD 2009 (2009)
29. Tsourakakis, C.E., Kolountzakis, M.N., Miller, G.L.: Triangle sparsifiers. J. Graph Algorithms Appl.

30. Watts, D.J., Strogatz, S.H.: Collective dynamics of small-world networks. Nature
31. Zhang, B., Horvath, S., et al.: A general framework for weighted gene co-expression
    network analysis. Statistical Applications in Genetics and Molecular Biology
32. Zhang, Y., Zhang, Z., Guan, J., Zhou, S.: Analytic solution to clustering coefficients
    on weighted networks. arXiv preprint arXiv:0911.0476 (2009)

Global Clustering Coefficient in Scale-Free Networks

Liudmila Ostroumova Prokhorenkova[1,2,3]([⊠]) and Egor Samosvat[1,3]

[1] Yandex, Moscow, Russia
ostroumova-la@yandex.ru
[2] Moscow State University, Moscow, Russia
[3] Moscow Institute of Physics and Technology, Moscow, Russia
samosvat.egor@gmail.com

Abstract. In this paper, we analyze the behavior of the global clustering coefficient in scale free graphs. We are especially interested in the case of degree distribution with an infinite variance, since such degree distribution is usually observed in real-world networks of diverse nature.

There are two common definitions of the clustering coefficient of a graph: global clustering and average local clustering. It is widely believed that in real networks both clustering coefficients tend to some positive constant as the networks grow. There are several models for which the average local clustering coefficient tends to a positive constant. On the other hand, there are no models of scale-free networks with an infinite variance of degree distribution and with a constant global clustering.

In this paper we prove that if the degree distribution obeys the power law with an infinite variance, then the global clustering coefficient tends to zero with high probability as the size of a graph grows.

1 Introduction

In this paper, we analyze the global clustering coefficient of graphs with a power-law degree distribution. Namely, we consider a sequence of graphs with the degree distribution following a regularly varying distribution F. Our main result is the following. If the degree distribution has an infinite variance, then the global clustering coefficient tends to zero with high probability.

It is important to note that we do not specify any random graph model, our result holds for any sequence of graphs. The only restriction we have on a sequence is the distribution of degrees: we assume that the degrees of vertices (except one vertex, see the explanation at the end of Section 3) are i.i.d. random variables following a regularly varying distribution with a parameter $1 < \gamma < 2$.

Our results are especially interesting taking into account the fact that it was suspected that for many types of networks both the average local and the global clustering coefficients tend to non-zero limit as the network becomes large. It is a natural assumption as in many observed networks the values of both clustering coefficients are considerably high [9]. Note that actually these observations do not contradict ours:

© Springer International Publishing Switzerland 2014
A. Bonato et al. (Eds.): WAW 2014, LNCS 8882, pp. 47–58, 2014.
DOI: 10.1007/978-3-319-13123-8_5

- Large values of global clustering coefficient are usually obtained on small networks.
- For the networks with the power-law degree distribution the observed global clustering is usually less than the average local clustering, as expected.
- Our results can be applied only to networks with regularly varying degree distribution. If a network has, for example, a power-law degree distribution with an exponential cut-off, then our results cannot be applied.

The rest of the paper is organized as follows. In the next section, we discuss two definitions of the clustering coefficient. Then, in Section 3, we formally define our restriction on the sequence of graphs. In Section 4, we prove that a simple graph with the given degree sequence exists with high probability. In Section 5, we prove that the global clustering coefficient for any such sequence of graphs tends to zero. Then we discuss one graph constructing procedure which gives a sequence of graphs with superlinear number of triangles, but the global clustering coefficient for such sequence still tends to zero. Section 7 concludes the paper.

2   Clustering Coefficients

There are two popular definitions of the clustering coefficient [3,9]. The *global clustering coefficient* $C_1(G_n)$ is the ratio of three times the number of triangles to the number of pairs of adjacent edges in G_n. The *average local clustering coefficient* is defined as follows: $C_2(G_n) = \frac{1}{n} \sum_{i=1}^{n} C(i)$, where $C(i)$ is the local clustering coefficient for a vertex i: $C(i) = \frac{T^i}{P_2^i}$, where T^i is the number of edges between the neighbors of the vertex i and P_2^i is the number of pairs of neighbors. Note that both clustering coefficients equal 1 for a complete graph.

It was mentioned in [3,9] that in research papers either average local or global clustering are considered. And it is not always clear which definition is used. On the other hand, these two clustering coefficients differ. It was demonstrated in [11] that for networks based on the idea of preferential attachment the difference between these two clustering coefficients is crucial.

Note that both definitions of the clustering coefficient work only for graphs without multiple edges. Also, most measurements on real-world networks do not take multiple edges into account. Therefore, further we consider only simple graphs: graphs without loops and multiple edges. Clustering coefficient for weighted graphs can also be defined (see, e.g., [10]). We leave the analysis of the clustering coefficient in weighted graphs for the future work

3   Scale-Free Graphs

We consider a sequence of graphs $\{G_n\}$. Each graph G_n has n vertices. We assume that the degrees of these vertices are independent random variables following a *regularly varying* distribution with a cumulative distribution function F such that:

$$1 - F(x) = L(x)x^{-\gamma}, \quad x > 0, \tag{1}$$

where $L(\cdot)$ is a slowly varying function, that is, for any fixed constant $t > 0$

$$\lim_{x \to \infty} \frac{L(tx)}{L(x)} = 1.$$

There are other obvious restrictions on the function $L(\cdot)$, for instance, the function $1 - L(x)x^{-\gamma}$ must be a cumulative distribution function of a random variable taking positive integer values with probability 1. Further in this paper we use the following property of slowly varying functions: $L(x) = o(x^c)$ for any $c > 0$.

Note that (1) describes a broad class of heavy-tailed distributions without imposing the rigid Pareto assumption. Power-law distribution with parameter $\gamma + 1$ corresponds to the cumulative distribution $1 - F(x) = L(x)x^{-\gamma}$. Further by $\xi, \xi_1, \xi_2, \ldots$ we denote random variables with the distribution F. Note that for any $\alpha < \gamma$ the moment $\mathbb{E}\xi^\alpha$ is finite.

Models with $\gamma > 2$ and with the global clustering coefficient tending to some positive constant were already proposed (see, e.g., [11]). Therefore, in this paper we consider only the case $1 < \gamma < 2$.

One small problem remains: we can construct a graph with a given degree distribution only if the sum of degrees is even. This problem is easy to solve: we can either regenerate the degrees until their sum is even or we can add 1 to the last variable if their sum is odd [4]. For simplicity we choose the second option, i.e., if $\sum_{i=1}^n \xi_i$ is odd, then we replace ξ_n by $\xi_n + 1$. It is easy to see that this correction does not change any of our results, therefore, further we do not focus on the evenness.

4 Existence of a Graph with Given Degree Distribution

4.1 Result

As pointed out in [8], the probability of obtaining a simple graph with given degree distribution by random pairing of edges' endpoints (configuration model) converges to a strictly positive constant if the degree distribution has a finite second moment. In our case the second moment is infinite and it can be shown that the probability of obtaining a simple graph just by random pairing of edges' endpoints tends to zero with n.

However, we can prove that in this case a simple graph with a given degree distribution exists with high probability and it can be constructed, e.g., using Havel-Hakimi algorithm [6,7].

Theorem 1. *For any δ such that $1 < \delta < \gamma$ with probability $1 - O\left(n^{1-\delta}\right)$ there exists a simple graph on n vertices with the degree distribution defined above.*

4.2 Auxiliary Results

We use the following theorem proved by Erdős and Gallai in 1960 [5].

Theorem 2 (Erdős–Gallai). *A sequence of non-negative integers $d_1 \geq \ldots \geq d_n$ can be represented as the degree sequence of a finite simple graph on n vertices if and only if*

1. $d_1 + \ldots + d_n$ *is even;*
2. $\sum_{i=1}^{k} d_i \leq k(k-1) + \sum_{i=k+1}^{n} \min(d_i, k)$ *holds for $1 \leq k \leq n$.*

In this case a sequence $d_1 \geq \ldots \geq d_n$ is called *graphic*.

If a degree sequence is graphic, then one can use Havel-Hakimi algorithm to construct a simple graph corresponding to it [6,7]. The idea of the algorithm is the following. We sort degrees in nondecreasing order $d_1 \geq \ldots \geq d_n$. Then we take the vertex of the highest degree d_1 and connect this vertex to the vertices of degrees $d_2, \ldots, d_{d_1+1}$. After this we get the degree sequence $d_2 - 1, \ldots, d_{d_1+1} - 1, d_{d_1+2}, \ldots, d_n$ and apply the same procedure to this sequence, and so on.

We also use the following theorem several times in this paper (see, e.g., [1]).

Theorem 3 (Karamata's theorem). *Let L be slowly varying and locally bounded in $[x_0, \infty]$ for some $x_0 \geq 0$. Then*

1. *for $\alpha > -1$*

$$\int_{x_0}^{x} t^{\alpha} L(t) dt = (1 + o(1))(\alpha + 1)^{-1} x^{\alpha+1} L(x), \quad x \to \infty.$$

2. *for $\alpha < -1$*

$$\int_{x}^{\infty} t^{\alpha} L(t) dt = -(1 + o(1))(\alpha + 1)^{-1} x^{\alpha+1} L(x), \quad x \to \infty.$$

We also use the following known lemma (proof can be found, e.g., in [12]).

Lemma 1. *Let $\xi_1, \ldots, \xi_n$ be mutually independent random variables, $\mathbb{E}\xi_i = 0$, $\mathbb{E}|\xi_i|^{\alpha} < \infty$, $1 \leq \alpha \leq 2$, then*

$$\mathbb{E}\left[|\xi_1 + \ldots + \xi_n|^{\alpha}\right] \leq 2^{\alpha} \left(\mathbb{E}\left[|\xi_1|^{\alpha}\right] + \ldots + \mathbb{E}\left[|\xi_n|^{\alpha}\right]\right).$$

4.3 Proof of Theorem 1

We need the following lemma on the number of edges in the graph.

Lemma 2. *For any θ such that $1 < \theta < \gamma$ with probability $1 - O(n^{1-\theta})$ the number of edges $E(G_n)$ in our graph satisfies the following inequalities:*

$$\frac{n\mathbb{E}\xi}{4} \leq E(G_n) \leq \frac{3n\mathbb{E}\xi}{4}.$$

Proof. The expectation of the number of edges is

$$\mathbb{E}E(G_n) = n\mathbb{E}\xi/2\,.$$

Note that for $1 < \theta < \gamma$ we have $\mathbb{E}|\xi - \mathbb{E}\xi|^\theta < \infty$ and

$$\mathsf{P}\left(|E(G_n) - n\mathbb{E}\xi/2| \geq n\mathbb{E}\xi/4\right) \leq \frac{4^\theta \mathbb{E}\left|\sum_{i=1}^n (\xi_i - \mathbb{E}\xi)/2\right|^\theta}{n^\theta (\mathbb{E}\xi)^\theta}$$

$$\leq \frac{8^\theta n \mathbb{E}|\xi - \mathbb{E}\xi|^\theta}{n^\theta (\mathbb{E}\xi)^\theta} = O\left(n^{1-\theta}\right).$$

Here we applied Lemma 1. This concludes the proof of Lemma 2.

Let us order the random variables $\xi_1, \ldots, \xi_n$ and obtain the ordered sequence $d_1 \geq \ldots \geq d_n$.

We want to show that with probability $1 - O\left(n^{1-\delta}\right)$ the condition

$$\sum_{i=1}^k d_i \leq k(k-1) + \sum_{i=k+1}^n \min(d_i, k) \tag{2}$$

holds for all k, $1 \leq k \leq n$.

Note that if $k \geq \sqrt{2\mathbb{E}\xi n}$, then with probability $1 - O\left(n^{1-\delta}\right)$ we have

$$\sum_{i=1}^k d_i \leq 2E(G_n) \leq k(k-1)$$

as $2E(G_n) \leq \frac{3\mathbb{E}\xi}{2}n$ (here we apply Lemma 2 with $\theta = \delta$). Therefore the condition (2) is satisfied.

Now consider the case $k < \sqrt{2\mathbb{E}\xi n}$. In this case we will show that

$$\sum_{i=k+1}^n \min(d_i, k) \geq \sum_{i=1}^k d_i$$

which implies the condition (2). Note that $\min(d_i, k) > 1$ so

$$\sum_{i=k+1}^n \min(d_i, k) \geq n - \sqrt{2\mathbb{E}\xi n}\,.$$

It remains to show that with probability $1 - O\left(n^{1-\delta}\right)$ we have

$$\sum_{i=1}^{\sqrt{2\mathbb{E}\xi n}} d_i \leq n - \sqrt{2\mathbb{E}\xi n}\,. \tag{3}$$

Fix some α such that $\delta < \alpha < \gamma$. Consider any β such that

$$0 < \beta < \min\left\{\frac{2-\delta}{\gamma}, \frac{1}{2\gamma}, \frac{\alpha-\delta}{\gamma(\alpha-1)}\right\} \tag{4}$$

and let

$$S_n = \sum_{i=1}^{n} \xi_i I\left[\xi_i > n^\beta\right].$$

We will show that with probability $1 - O\left(n^{1-\delta}\right)$ we have

$$\sum_{i=1}^{\sqrt{2\mathbb{E}\xi n}} d_i \le S_n \le n - \sqrt{2\mathbb{E}\xi n} \tag{5}$$

which implies (3). Note that in order to prove the left inequality it is sufficient to show that with probability $1 - O\left(n^{1-\delta}\right)$ we have

$$S_n' := \sum_{i=1}^{n} I\left[\xi_i > n^\beta\right] \ge \sqrt{2\mathbb{E}\xi n}.$$

The expectation of S_n' is

$$\mathbb{E}S_n' = \mathbb{E}\sum_{i=1}^{n} I\left[\xi_i > n^\beta\right] = n\mathsf{P}\left(\xi > n^\beta\right) = nL\left(n^\beta\right)n^{-\gamma\beta}.$$

Now we will show the concentration:

$$\mathsf{P}\left(|S_n' - \mathbb{E}S_n'| > \frac{\mathbb{E}S_n'}{2}\right) \le \frac{4\mathrm{Var}(S_n')}{(\mathbb{E}S_n')^2} =$$

$$= \frac{4n\left(L\left(n^\beta\right)n^{-\gamma\beta} - \left(L\left(n^\beta\right)\right)^2 n^{-2\gamma\beta}\right)}{n^2\left(L\left(n^\beta\right)\right)^2 n^{-2\gamma\beta}} = O\left(\frac{n^{\gamma\beta}}{nL\left(n^\beta\right)}\right) = O\left(n^{1-\delta}\right).$$

Here in the last equation we use the inequality $\beta < \frac{2-\delta}{\gamma}$, so $\gamma\beta - 1 < 1 - \delta$. It remains to note that as $\beta < \frac{1}{2\gamma}$ for large enough n we have

$$\frac{1}{2}nL\left(n^\beta\right)n^{-\gamma\beta} \ge \sqrt{2\mathbb{E}\xi n}.$$

Now let us prove the right inequality in (5), i.e., prove that with probability $1 - O\left(n^{1-\delta}\right)$ we have

$$S_n \le n - \sqrt{2\mathbb{E}\xi n}.$$

As before, first we estimate the expectation of S_n:

$$\mathbb{E}S_n = n\int_{n^\beta}^{\infty} x\,dF(x) = -n\int_{n^\beta}^{\infty} x\,d(1 - F(x))$$

$$= -n\,x(1 - F(x))\Big|_{n^\beta}^{\infty} + n\int_{n^\beta}^{\infty} (1 - F(x))\,dx$$

$$= n\,n^\beta n^{-\gamma\beta}L\left(n^\beta\right) + n\int_{n^\beta}^{\infty} x^{-\gamma}L(x)\,dx$$

$$\sim n^{1+\beta(1-\gamma)}L\left(n^\beta\right) + n(\gamma - 1)^{-1}n^{\beta(1-\gamma)}L\left(n^\beta\right) = \frac{\gamma}{\gamma - 1}n^{1+\beta(1-\gamma)}L\left(n^\beta\right).$$

In order to show concentration we first estimate

$$\mathbb{E}\left(\xi I\left[\xi > n^{\beta}\right]\right)^{\alpha} = -\int_{n^{\beta}}^{\infty} x^{\alpha}\, d(1 - F(x))$$

$$= -x^{\alpha}(1 - F(x))\Big|_{n^{\beta}}^{\infty} + \int_{n^{\beta}}^{\infty}(1 - F(x))\, dx^{\alpha}$$

$$= n^{\alpha\beta} n^{-\gamma\beta} L\left(n^{\beta}\right) + \alpha \int_{n^{\beta}}^{\infty} x^{\alpha-\gamma-1} L(x)\, dx$$

$$\sim n^{\beta(\alpha-\gamma)} L\left(n^{\beta}\right) + (\gamma - \alpha)^{-1} n^{\beta(\alpha-\gamma)} L\left(n^{\beta}\right) = \frac{\gamma + 1 - \alpha}{\gamma - 1} n^{\beta(\alpha-\gamma)} L\left(n^{\beta}\right).$$

We get

$$\mathsf{P}\left(|S_n - \mathbb{E}S_n| > \frac{\mathbb{E}S_n}{2}\right) \leq \frac{\mathbb{E}|S_n - \mathbb{E}S_n|^{\alpha}}{(\mathbb{E}S_n)^{\alpha}}$$

$$= O\left(\frac{n\mathbb{E}\left(\xi I\left[\xi > n^{\beta}\right]\right)^{\alpha}}{(\mathbb{E}S_n)^{\alpha}}\right) = O\left(\frac{n^{1+\beta(\alpha-\gamma)} L\left(n^{\beta}\right)}{n^{\alpha(1+\beta(1-\gamma))}\left(L\left(n^{\beta}\right)\right)^{\alpha}}\right) = O\left(n^{1-\delta}\right).$$

Here in the last equation we use the inequality $\beta < \frac{\alpha-\delta}{\gamma(\alpha-1)}$.

It remains to note that as $\gamma > 1$ for large n we have

$$\frac{\gamma}{2(\gamma - 1)} n^{1+\beta(1-\gamma)} L\left(n^{\beta}\right) < n - \sqrt{2\mathbb{E}\xi n}.$$

5   Global Clustering Coefficient

5.1   Result

Theorem 4. *For any $\varepsilon > 0$ and any α such that $1 < \alpha < \gamma$ with probability $1 - O(n^{1-\alpha})$ the global clustering coefficient satisfies the following inequality*

$$C_1(G_n) \leq n^{\varepsilon - \frac{(\gamma-2)^2}{2\gamma}}.$$

Taking small enough ε one can see that with high probability $C_1(G_n) \to 0$ as n grows.

5.2   Proof of Theorem 4

We will use the following estimate for $C_1(G_n)$:

$$C_1(G_n) \leq \frac{E(G_n)\left|\{i : \xi_i^2 \geq E(G_n)\}\right| + \sum_{i:\xi_i^2 < E(G_n)} \xi_i^2}{P_2(n)}. \tag{6}$$

Here $P_2(n)$ is the number of pairs of adjacent edges in G_n. In order to obtain inequality (6) we use the following observation. The number of triangles connected to a vertex cannot be larger than the number of edges in a graph. Also, this number cannot be larger than the degree squared.

Using Lemma 2 (with $\theta = \alpha$) we get that with probability $1 - O\left(n^{1-\alpha}\right)$

$$C_1(G_n) \leq \frac{\frac{3\mathbb{E}\xi n}{4}\left|\left\{i : \xi_i^2 \geq \frac{\mathbb{E}\xi n}{4}\right\}\right| + \sum_{i:\xi_i^2 < \frac{3\mathbb{E}\xi n}{4}} \xi_i^2}{P_2(n)}. \tag{7}$$

Let us find a lower bound for the number of pairs of adjacent edges $P_2(n)$.

Lemma 3. *For any $\delta > 0$ and any α such that $1 < \alpha < \gamma$ with probability $1 - O\left(n^{1-\alpha}\right)$ we have*

$$P_2(n) \geq n^{\frac{2}{\gamma}-\delta}.$$

Proof. Let $\xi_{max} = \max\{\xi_1, \ldots, \xi_n\}$, then

$$P_2(n) \geq \frac{\xi_{max}(\xi_{max} - 1)}{2}.$$

It remains to find a lower bound for ξ_{max} now:

$$P(\xi_{max} < 2n^{\frac{1}{\gamma}-\frac{\delta}{2}}) = \left[P\left(\xi < 2n^{\frac{1}{\gamma}-\frac{\delta}{2}}\right)\right]^n = \exp\left(n\log\left(1 - P(\xi \geq 2n^{\frac{1}{\gamma}-\frac{\delta}{2}})\right)\right)$$

$$= \exp\left(n\log\left(1 - L\left(2n^{\frac{1}{\gamma}-\frac{\delta}{2}}\right)2^{-\gamma}n^{-\gamma\left(\frac{1}{\gamma}-\frac{\delta}{2}\right)}\right)\right)$$

$$= \exp\left(-n\left(L\left(2n^{\frac{1}{\gamma}-\frac{\delta}{2}}\right)2^{-\gamma}n^{-1+\gamma\frac{\delta}{2}}\right)(1+o(1))\right)$$

$$= \exp\left(-L\left(2n^{\frac{1}{\gamma}-\frac{\delta}{2}}\right)2^{-\gamma}n^{\gamma\frac{\delta}{2}}(1+o(1))\right) = O\left(n^{1-\alpha}\right).$$

So, with probability $1 - O\left(n^{1-\alpha}\right)$ we have

$$P_2(n) \geq n^{\frac{2}{\gamma}-\delta}.$$

This concludes the proof of Lemma 3

Now we estimate the number of vertices with large degrees.

Lemma 4. *For any $\delta > 0$ and any α such that $1 < \alpha < \gamma$ we have*

$$P\left(\left|\left\{i : \xi_i \geq \sqrt{\frac{\mathbb{E}\xi n}{4}}\right\}\right| \leq n^{1-\frac{\gamma}{2}+\delta}\right) = 1 - O\left(n^{1-\alpha}\right).$$

Proof. Let

$$S_n' := \sum_{i=1}^{n} I\left[\xi_i \geq \sqrt{\frac{\mathbb{E}\xi n}{4}}\right].$$

The expectation of S_n' is

$$\mathbb{E}S_n' = \mathbb{E}\sum_{i=1}^{n} I\left[\xi_i \geq \sqrt{\frac{\mathbb{E}\xi n}{4}}\right] = n P\left(\xi \geq \sqrt{\frac{\mathbb{E}\xi n}{4}}\right)$$

$$= n\left(\frac{\mathbb{E}\xi n}{4}\right)^{-\gamma/2} L\left(\sqrt{\frac{\mathbb{E}\xi n}{4}}\right).$$

We can apply Hoeffding's inequality:

$$P\left(S_n' > 2\mathbb{E}S_n'\right) \le \exp\left(-2(\mathbb{E}S_n')^2 n\right) = O\left(n^{1-\alpha}\right).$$

It remains to note that for large enough n we have

$$2\mathbb{E}S_n' < n^{1-\frac{7}{2}+\delta}.$$

Lemma 5. *For any $\delta > 0$ and any α such that $1 < \alpha < \gamma$ we have*

$$P\left(\sum_{i:\xi_i^2 < \frac{3\mathbb{E}\xi n}{4}} \xi_i^2 \le n^{2-\gamma/2+\delta}\right) = 1 - O\left(n^{1-\alpha}\right).$$

Proof. Let

$$S_n = \sum_{i=1}^n \xi_i^2 I\left[\xi_i < \sqrt{\frac{3\mathbb{E}\xi n}{4}}\right].$$

Again, we first estimate the expectation of S_n. Since $L(x)$ is locally bounded we can apply Karamata's theorem:

$$\mathbb{E}S_n = -n \int_0^{\sqrt{\frac{3\mathbb{E}\xi n}{4}}} x^2 d(1 - F(x))$$

$$= -n\,x^2(1 - F(x))\Big|_0^{\sqrt{\frac{3\mathbb{E}\xi n}{4}}} + 2n \int_0^{\sqrt{\frac{3\mathbb{E}\xi n}{4}}} x(1 - F(x))\,dx$$

$$= n\left(\frac{3\mathbb{E}\xi n}{4}\right)^{1-\gamma/2} L\left(\sqrt{\frac{3\mathbb{E}\xi n}{4}}\right) + 2n \int_0^{\sqrt{\frac{3\mathbb{E}\xi n}{4}}} x^{1-\gamma} L(x)\,dx$$

$$\sim n\left(\frac{3\mathbb{E}\xi n}{4}\right)^{1-\gamma/2} L\left(\sqrt{n}\right) + 2n(2-\gamma)^{-1}\left(\frac{3\mathbb{E}\xi n}{4}\right)^{1-\gamma/2} L\left(\sqrt{n}\right)$$

$$= \frac{4-\gamma}{2-\gamma}\left(\frac{3\mathbb{E}\xi}{4}\right)^{1-\gamma/2} n^{2-\gamma/2} L\left(\sqrt{n}\right).$$

In order to show concentration we first estimate

$$\mathbb{E}\left(\xi^2 I\left[\xi < \sqrt{\frac{3\mathbb{E}\xi n}{4}}\right]\right)^\alpha = -\int_0^{\sqrt{\frac{3\mathbb{E}\xi n}{4}}} x^{2\alpha} d(1 - F(x))$$

$$= -x^{2\alpha}(1 - F(x))\Big|_0^{\sqrt{\frac{3\mathbb{E}\xi n}{4}}} x^{2\alpha} + \int_0^{\sqrt{\frac{3\mathbb{E}\xi n}{4}}} (1 - F(x))\,dx^{2\alpha}$$

$$= \left(\frac{3\mathbb{E}\xi n}{4}\right)^{\alpha-\gamma/2} L\left(\sqrt{\frac{3\mathbb{E}\xi n}{4}}\right) + 2\alpha \int_0^{\sqrt{\frac{3\mathbb{E}\xi n}{4}}} x^{2\alpha-\gamma-1} L(x)\,dx$$

$$\sim \left(\frac{3\mathbb{E}\xi n}{4}\right)^{\alpha-\gamma/2} L\left(\sqrt{n}\right) + (2\alpha - \gamma)^{-1}\left(\frac{3\mathbb{E}\xi n}{4}\right)^{\alpha-\gamma/2} L\left(\sqrt{n}\right)$$

$$= O\left(n^{\alpha-\gamma/2} L\left(\sqrt{n}\right)\right).$$

And we get

$$
\mathbb{P}\left(|S_n - \mathbb{E}S_n| > \frac{\mathbb{E}S_n}{2}\right) \leq \frac{\mathbb{E}|S_n - \mathbb{E}S_n|^\alpha}{(\mathbb{E}S_n)^\alpha}
$$

$$
\leq \frac{2^\alpha n \mathbb{E}\left(\xi^2 I\left[\xi < \sqrt{\frac{3\mathbb{E}\xi n}{4}}\right]\right)^\alpha}{(\mathbb{E}S_n)^\alpha} = O\left(\frac{n^{\alpha-\gamma/2}L\left(\sqrt{n}\right)}{n^{\alpha(2-\gamma/2)}\left(L\left(\sqrt{n}\right)\right)^\alpha}\right) = O\left(n^{1-\alpha}\right).
$$

Here in the last equation we use the inequality $\alpha < 2 < \frac{2}{\gamma-1}$.
It remains to note that for large enough n we have

$$
\frac{3(4-\gamma)}{2(2-\gamma)}\left(\frac{3\mathbb{E}\xi}{4}\right)^{1-\gamma/2} n^{2-\gamma/2}L\left(\sqrt{n}\right) \leq n^{2-\gamma/2+\delta}.
$$

This concludes the proof of Lemma 5

Theorem 4 follows immediately from Lemmas 3, 4, 5, and Equation (7).

6   Experiments

In the previous section, we proved that for any sequence of graphs with a regularly varying degree distribution with a parameter $1 < \gamma < 2$ the global clustering coefficient tends to zero at least as fast as $n^{-\frac{(\gamma-2)^2}{2\gamma}}$. In this case the number of pairs of adjacent edges is superlinear in the number of vertices and it grows faster than the number of triangles.

In this section, we present a simple method which allows to construct scale-free graphs with a superlinear number of triangles. Consider a sequence of graphs constructed according to Havel-Hakimi algorithm. On Figure 1 we present the number of triangles, the number of pairs of adjacent edges, and the global clustering coefficient for such graphs. For each n we averaged the results over 100 independent samples of power-law degree distribution. Note that for $\gamma > 2$ the number of pairs of adjacent edges grows linearly and for $1 < \gamma < 2$ it grows as $n^{2/\gamma}$, as expected. The number of triangles grows linearly for $\gamma > 2$ and grows as $n^{3/(\gamma+1)}$ for $1 < \gamma < 2$. The constant $3/(\gamma+1)$ can be explained in the following way. If the degree distribution follows the power law with a parameter γ, then the maximum clique which can be obtained is of size $n^{\frac{1}{\gamma+1}}$ since $d_k \approx k$ for $k \sim n^{\frac{1}{\gamma+1}}$. This clique gives $\binom{k}{3}$ triangles. Since Havel-Hakimi algorithm also connects the vertices of largest degrees to each other, we get $\sim n^{3/(\gamma+1)}$ triangles.

To sum up, we can construct a sequence of graphs with $n^{\frac{3}{\gamma+1}}$ triangles and our theoretical upper bound is $n^{2-\frac{\gamma}{2}}$. It is easy to see that for $1 < \gamma < 2$ we have $\frac{3}{\gamma+1} < 2 - \frac{\gamma}{2}$. So, there is a gap between the number of constructed triangles and the upper bound.

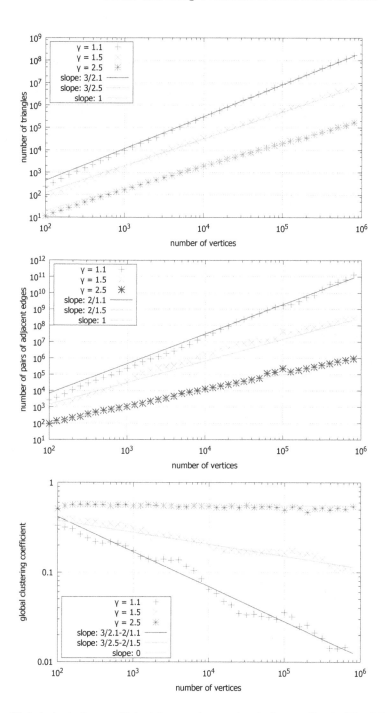

Fig. 1. Global clustering coefficient for graphs constructed according to Havel-Hakimi algorithm

7 Conclusion

In this paper, we analyzed the global clustering coefficient in scale-free graphs. We proved that for any sequence of graphs with a regularly varying degree distribution with a parameter $1 < \gamma < 2$ the global clustering coefficient tends to zero with high probability. We also proved that with high probability a graph with the required degree distribution exists.

Finally, we demonstrated the construction procedure which allows to obtain the sequence of graphs with superlinear number of triangles. Unfortunately, the number of triangles in this case grows slower than the upper bound obtained in Section 5.

References

1. Bingham, N.H., Goldie, C.M., Teugels, J.L.: Regular Variation. Cambridge University Press, Cambridge (1987)
2. Boccaletti, S., Latora, V., Moreno, Y., Chavez, M., Hwang, D.-U.: Complex networks: Structure and dynamics. Physics Reports **424**(45), 175–308 (2006)
3. Bollobás, B., Riordan, O.M.: Mathematical results on scale-free random graphs. In: Handbook of Graphs and Networks: From the Genome to the Internet, pp. 1–3 (2003)
4. Britton, T., Deijfen, M., Martin-Löf, A.: Generating simple random graphs with prescribed degree distribution. J. Stat. Phys. **124**(6), 1377–1397 (2006)
5. Erdős, P., Gallai, T.: Graphs with given degrees of vertices. Mat. Lapok **11**, 264–274 (1960)
6. Hakimi, S.: On the realizability of a set of integers as degrees of the vertices of a graph. SIAM Journal of Applied Mathematics **10**, 496–506 (1962)
7. Havel, V.: A remark on the existence of finite graphs [In Czech.]. Casopis Pro Pestovani Matematiky **80**, 477–480 (1955)
8. Molloy, M., Reed, B.: A critical point for random graphs with a given degree sequence. Rand. Struct. Alg. **6**, 161–179 (1995)
9. Newman, M.E.J.: The structure and function of complex networks. SIAM Review **45**, 167–256 (2003)
10. Opsahl, T., Panzarasa, P.: Clustering in weighted networks. Social Networks **31**(2), 155–163 (2009)
11. Ostroumova, L., Ryabchenko, A., Samosvat, E.: Generalized Preferential Attachment: Tunable Power-Law Degree Distribution and Clustering Coefficient. In: Bonato, A., Mitzenmacher, M., Prałat, P. (eds.) WAW 2013. LNCS, vol. 8305, pp. 185–202. Springer, Heidelberg (2013)
12. Ostroumova, L., Samosvat, E.: Recency-based preferential attachment models (2014). http://arxiv.org/abs/1406.4308

Efficient Primal-Dual Graph Algorithms for MapReduce

Bahman Bahmani[1]([✉]), Ashish Goel[2], and Kamesh Munagala[3]

[1] Department of Computer Science, Stanford University, Stanford, USA
bahman@cs.stanford.edu
[2] Department of Management Science and Engineering, Stanford University, Stanford, USA
ashishg@stanford.edu
[3] Department of Computer Science, Duke University, Durham, USA
kamesh@cs.duke.edu

Abstract. In this paper, we obtain improved algorithms for two graph-theoretic problems in the popular MAPREDUCE framework. The first problem we consider is the densest subgraph problem. We present a primal-dual algorithm that provides a $(1 + \epsilon)$ approximation and takes $O(\frac{\log n}{\epsilon^2})$ MAPREDUCE iterations, each iteration having a shuffle size of $O(m)$ and a reducer size of $O(d_{max})$. Here m is the number of edges, n is the number of vertices, and d_{max} is the maximum degree of a node. This dominates the previous best MAPREDUCE algorithm, which provided a $(2 + \delta)$-approximation in $O(\frac{\log n}{\delta})$ iterations, with each iteration having a total shuffle size of $O(m)$ and a reducer size of $O(d_{max})$.

The standard primal-dual technique for solving the above problem results in $O(n)$ iterations. Our key idea is to carefully control the width of the underlying polytope so that the number of iterations becomes small, but an approximate primal solution can still be recovered from the approximate dual solution. We then show an application of the same technique to the fractional maximum matching problem in bipartite graphs. Our results also map naturally to the PRAM model.

1 Introduction

Over the last two decades, the primal-dual method (e.g., [3,9,14,17]) has been used to solve many graph optimization problems. Recently, the programming paradigm MAPREDUCE [7] and its main open-source implementation, Hadoop [10], have had an enormous impact on large scale data processing. In this paper, our goal

Ashish Goel: Supported in part by the DARPA xdata program, by grant #FA9550-12-1-0411 from the U.S. Air Force Office of Scientific Research (AFOSR) and the Defense Advanced Research Projects Agency (DARPA), and by NSF Award 0904325.

Kamesh Munagala: Supported by NSF grants CCF- 0745761, CCF-1348696, IIS-0964560, and IIS-1447554; and by grant W911NF-14-1- 0366 from the Army Research Office (ARO). Part of this work was done while the author was visiting Twitter, Inc.

© Springer International Publishing Switzerland 2014
A. Bonato et al. (Eds.): WAW 2014, LNCS 8882, pp. 59–78, 2014.
DOI: 10.1007/978-3-319-13123-8_6

is to adapt the primal-dual technique for solving graph optimization problems to the MAPREDUCE framework. We consider the densest subgraph problem as well as the fractional maximum bipartite matching problem, both of which have the following structure: These problems can be written as linear programs; furthermore, either the linear program or its dual can be interpreted as solving maximum or concurrently maximum flows on suitably defined bipartite graphs. We present a simple algorithmic technique for solving these optimization problems to arbitrarily good approximations, which yields fast sequential as well as MAPREDUCE algorithms, thus improving the state of the art.

1.1   Problem Formulations and Results

We will focus on two graph optimization problems in this paper (although our technique is more widely applicable). The first is the classic densest subgraph problem in general graphs. This can be defined either for undirected or directed graphs. For an undirected graph $G(V, E)$ with n vertices and m edges, the problem is defined as follows. Find a subset H of vertices such that the induced subgraph $G'(H, F)$ has maximum density (or average degree), $|F|/|H|$. For a directed graph $G(V, E)$, the densest subgraph problem [11] asks to find two subsets $S, T \subseteq V$ (not necessarily disjoint) such that if $E(S, T)$ denotes the set of edges (u, v) where $u \in S$ and $v \in T$, then the density $D(S, T) = \frac{|E(S,T)|}{\sqrt{|S||T|}}$ is maximized. This problem has various applications for web, social, and biological networks, such as identifying spammers, dense communities of users, or functional genetic modules. We refer the reader to [5, 15] for several applications.

The other problem we consider is the maximum generalized fractional matchings on bipartite graphs. This problem has several applications in Adword allocation and load balancing [13]. Specifically, we consider the following allocation problem [13]: There are n_1 advertisers, where advertiser i has demand d_i; there are n_2 ad slots, where slot j has supply s_j; furthermore there is a bipartite graph specifying which advertiser is interested in which ad slots. The goal is to find an allocation of slots to advertisers that maximizes the demand satisfied; if the supply and demands are large, this can be thought of as a fractional allocation without loss of generality.

MapReduce Model. We base the presentation of our MAPREDUCE results on the model described in [1], which takes into account practical aspects of MAPRE-DUCE, such as shuffle costs and the "curse of the last reducer" [16], while still being independent of the implementation details such as number of processors, Mappers, and Reducers (unlike abstractions such as in [12]). For a phase of a MAPREDUCE computation, the model in [1] has two measures, reducer size and replication rate. Reducer size is the maximum number of values associated with a reduce key and captures the maximum input size, maximum memory, and maximum running time for a single reduce key per phase. Replication rate is the average number of (key; value) pairs produced per input map key per phase. For the sake of simplicity of presentation, instead of replication rate, we will use an equivalent measure that we denote as sequential complexity, which captures the total output size (and hence, the complexity of the shuffle operation, ignoring implementation details) and total running time of all mappers and reducers in a phase. Note that in this paper, the mappers and reducers are linear (or near-linear) time and space algorithms. Therefore, the distinction between input size, memory, and running time is not important. Hence, we will express our MAPRE-DUCE results using three measures, namely, reducer size, sequential complexity, and number of phases.

Our Results. In this paper, we develop a *simple and general algorithmic technique* that achieves a $(1 + \epsilon)$ approximation to the above problems (for any $\epsilon > 0$), with sequential running times of the form $O(mf(\log n, 1/\epsilon))$ where f is a polynomial. Here, m is the number of edges, and n the number of vertices in the input graph. We will be interested in making the function f as small as possible. We will further design MAPREDUCE implementations that minimize the number of phases subject to two constraints: (1) The *reducer size* (i.e. the maximum size/computation/memory associated with any map or reduce key) is $O(d_{\max})$, the maximum degree, and (2) The *sequential complexity* (the total volume of data transferred, or the total time spent) in any phase is $\tilde{O}(m)$, where the $\tilde{O}$ notation hides low order terms ($\frac{\log n}{\epsilon}$ to be precise). Our results are summarized as follows, where the running time refers to sequential running times.

- UNDIRECTED DENSEST SUBGRAPH: We present an algorithm that runs in $O\left(m\frac{\log n}{\epsilon^2}\right)$ time, and takes $O\left(\frac{\log n}{\epsilon^2}\right)$ MAPREDUCE phases (Section 2). The total running time and shuffle size in any one phase is $O(m)$. The best previous MAPREDUCE implementation is a greedy algorithm in [5]; this algorithm takes $O\left(\frac{\log n}{\delta}\right)$ phases to yield a $2 + \delta$ approximation[1].

- DIRECTED DENSEST SUBGRAPH: We combine the above approach with a linear programming formulation due to Charikar [6] to obtain $\tilde{O}\left(m\frac{\log^2 n}{\epsilon^3}\right)$ running time, and $O\left(\frac{\log n}{\epsilon^2}\right)$ MAPREDUCE phases (Appendix B). Again, the best previous best MAPREDUCE algorithm [5] was a greedy $(2 + \delta)$ approximation.

- FRACTIONAL MATCHING on bipartite graphs: We show that exactly the same technique yields $\tilde{O}\left(m\frac{\log^2 n}{\epsilon^3}\right)$ running time, and $O\left(\frac{\log^2 n}{\epsilon^3 \log d_{\max}}\right)$ MAPREDUCE phases (Appendix C). This matches (and in fact, improves by a $\log d_{\max}$ factor) that of the natural MAPREDUCE implementation of the semi-streaming algorithm in [2][2]. Furthermore our sequential running time is comparable to the best sequential algorithms in [13] (which we do not know how to implement on MAPREDUCE). While the improvement in running time is small, note that the previous algorithms were tailored to the bipartite matching

---

[1] Superficially, the number of phases in [5] might seem smaller. However, note that even $\delta = 0$ for their work corresponds to $\epsilon = 1/2$ for us.

[2] Though Ahn and Guha [2] present several improvements to their basic algorithm, these are in the semi-streaming model and do not appear to produce corresponding improvements in the MAPREDUCE model.

problem, whereas our algorithm naturally follows from the more general framework we describe next.

While we have stated our results in the MAPREDUCE computation framework, they also map to (and are novel in) the PRAM framework: the reducer size corresponds to the parallel running time per phase (given n processors) and the sequential running time corresponds to the total amount of work done.

1.2 Technique: Width Modulation

We design our algorithms by exploiting connections to fast algorithms for packing and covering linear programs. We illustrate this for the densest subgraph problem. We start with a linear programming relaxation of the densest subgraph problem (due to Charikar [6]) and show that it is the dual of a maximum concurrent multi-commodity flow (MCMF) problem on a suitably defined bipartite graph (where the goal is to simultaneously route demand from *all* sources to a sink). We then proceed to exploit fast approximation algorithms for the MCMF problem due to Plotkin-Shmoys-Tardos [9,14] and Young [17], which are unified in a framework due to Arora-Hazan-Kale [3]. These algorithms suffer from two significant roadblocks.

First, the number of parallel iterations of these algorithms depends on the *width* of the problem, which in our case is proportional to the maximum degree $d_{\max}$. This dependence arises because the MCMF formulation attempts to route demands of size 1 concurrently in a graph with capacities either infinity or $d_{\max}$. The algorithms are based on the Lagrangian approach, which converts the MCMF problem to a maximum multi-commodity flow problem ignoring the demands, which can make the Lagrangian route demand that can be a factor $d_{\max}$ larger than the original demands. We overcome this hurdle by a technique that we term *width modulation*, whereby we add spurious capacity constraints to make capacities a small constant factor larger than the maximum demand, which is 1.

Though the above method reduces width to a constant and does not affect feasibility, this introduces the second roadblock: Adding capacities to the dual program changes the primal problem. Though this is not an issue for optimal solutions (after all, we did not affect feasibility), an *approximately* optimal primal solution for the changed problem need not yield an approximately optimal solution to the problem we intended to solve. We need the *modulation* part to overcome this hurdle – we show that for width being sufficiently large, we can indeed recover an approximately optimal primal solution via developing a problem-specific efficient *rounding scheme*. In a sense, there is a trade-off between running time (small width helps) and ability of the algorithm to recover the solution to the original problem (large width helps). We show that this trade-off leads to a final solution with constant width, which yields an $O\left(\frac{\log n}{\epsilon^2}\right)$ phase MAPREDUCE implementation and an $O\left(m\frac{\log n}{\epsilon^2}\right)$ overall running time.

For directed graphs, we use a parametric linear programming formulation due to Charikar [6]. To adapt this to our framework, we devise a parametric search procedure. We finally show that the same technique applies with small changes to the fractional bipartite matching problem, showing its generality.

In summary, we show that several linear programs which have wide practical applications can be efficiently approximated on MAPREDUCE by considering their dual formulation. We show that there is a tension between reducing the width of the dual problem (which is needed for efficiency) and recovering the primal solution itself, and present a general technique based on *width modulation* and *rounding* to achieve this trade-off without sacrificing efficiency or precision.

1.3 Related Work

In addition to presenting the LP formulations that we use, Charikar [6] also presents a greedy 2-approximation for the densest subgraph problem (see also [15]). This algorithm is modified to yield an efficient MAPREDUCE implementation in [5]; this algorithm takes $O\left(\frac{\log n}{\epsilon}\right)$ rounds to yield a $2 + \epsilon$ approximation.

There is a long line of work on fast approximate algorithms for covering linear programs; see [4] for a survey. One closely related line of work are the algorithms for *spreading metrics* due to Garg-Konemann [8] and their parallel implementation due to Awerbuch *et al.* [4]. These algorithms can possibly be applied to our dual formulations; however, when implemented in MAPREDUCE these methods will need poly$(\log n, 1/\epsilon)$ phases for a large degree polynomial, which is a far worse running time than what we show. For example, the algorithm of Awerbuch *et al.*, when applied to our setting, would result in $O(\frac{\log^6 n}{\epsilon^4})$ phases. Furthermore, note that the techniques in [4,8] can also be viewed as width reduction, where the width is reduced by adding constraints that the flow along a path is at most the minimum capacity on that path (see [3] for details). However, our width modulation technique is fundamentally different - we modulate the capacities themselves based on the demand being routed. In contrast with the technique in [4,8], our technique changes the description of the primal problem and does not preserve approximate optimality. Hence we need a problem-specific rounding scheme to recover the primal solution.

Roadmap. For lack of space, we relegate the description of the multiplicative weight method to Appendix A. We present the FPTAS and MAPREDUCE implementation for undirected densest subgraph in Section 2, that for directed densest subgraph in Appendix B, and for fractional bipartite matchings in Appendix C.

2 Undirected Densest Subgraph

For an undirected graph $G(V, E)$ with n vertices and m edges, the DENSEST SUBGRAPH problem is defined as follows. Find a subset $H \subseteq V$ of vertices such that the induced subgraph $G'(H, F)$ has maximum density (or average degree), $|F|/|H|$. We denote the optimum density by D^*. We present an algorithm that for any $\epsilon > 0$, outputs a subgraph of density $D^*(1 - \epsilon)$ in $\tilde{O}\left(m\frac{\log n}{\epsilon^2}\right)$ running time, where the $\tilde{O}(\cdot)$ notation ignores lower order terms. In the MAPREDUCE model, we show that with reducer size $O(d_{\max})$, where $d_{\max}$ is the maximum degree, and sequential complexity $O(m)$, the algorithm requires $O\left(\frac{\log n}{\epsilon^2}\right)$ phases.

2.1   Linear Program and Duality

Let D^* denote the optimal density. We will first present a well-known linear program which is known to compute the optimal solution [6]. For any vertex $v \in V$, let $x_v \in \{0, 1\}$ denote whether $v \in H$. For any edge $e \in E$, let $y_e \in \{0, 1\}$ denote whether $e \in F$, which are the edges induced by H. We relax x_v, y_e to be any real number. The value D^* is the solution to the following linear program:

$$\text{Maximize} \quad \sum_e y_e$$

$$y_e \leq x_v \ \forall e \in E, e \text{ incident on } v$$
$$\sum_v x_v \leq 1$$
$$x_v, y_e \geq 0 \quad \forall v \in V, e \in E$$

To interpret the above program, note that if $y_e = 1$ for $e = (u, v)$, then both x_u and x_v have to be 1. This implies the first constraint. The objective should maximize $(\sum_e y_e)/(\sum_v x_v)$. We can scale the values so that $\sum_v x_v = 1$, and enforce this as the second constraint. This means the objective now maximizes $\sum_e y_e$. Therefore, the value of the above LP is at least D^*. (It is in fact known that it is exactly D^*, but we will not need that fact.) For simplicity, we overload notation and denote the optimal value of the LP by D^*.

We now take the dual of the above program. Let α_{ev} denote the dual variable associated with the first constraint, and let D denote the dual variable associated with the second constraint. We parametrize the dual constraints by the variable D, and call these set of constraints DUAL(D):

$$\alpha_{eu} + \alpha_{ev} \geq 1 \ \forall e = (u, v) \in E$$
$$\sum_{e \text{ incident on } v} \alpha_{ev} \leq D \ \forall v \in V$$
$$\alpha_{ev} \geq 0 \ \forall e, v$$

Since the dual program is minimizing D, using strong duality, we have:

Lemma 1. DUAL(D) *is feasible iff* $D \geq D^*$.

We note that the dual program is a maximum concurrent multi commodity flow (MCMF) problem: Construct a bipartite directed graph $G'(U', V', E')$ as follows: $U' = E$, $V' = V$, and $E' = \{(e, v) \in E \times V | e \text{ is incident on } v\}$. Each node in U' has demand 1, and the nodes in V' are connected to a sink with directed edges of capacity D. The first constraint means that each demand of 1 is completely routed; the second constraint means that for all directed edges to the sink of capacity D, the flow routed is at most the capacity. Therefore, the goal is decide if all demand can be concurrently routed to the sink while satisfying the capacity constraints.

2.2   Width Modulation

We will apply the multiplicative weight update framework as described in Appendix A to decide feasibility of $\text{DUAL}(D)$ for given D. In particular, we will decide the feasibility of the set of constraints:

$$\alpha_{eu} + \alpha_{ev} \geq 1 \qquad \forall e = (u, v) \in E$$

subject to the polyhedral constraints $P(D)$ (which depends on parameter D):

$$\sum_{e \text{ incident on } v} \alpha_{ev} \leq D \ \forall v \in V$$
$$\alpha_{ev} \geq 0 \ \forall e \in E, v \in V$$

The dual vector corresponding to the constraint $\alpha_{eu} + \alpha_{ev} \geq 1$ is y_e, whose dimension is m. The main issue with a naive application of the method is the *width*. Each α_{ev} can be as large as D, so that the LHS of the above constraint can be as large as $2D$, which is also the width. Since D is the density and can be as large as n, this implies a polynomial number of MAPREDUCE rounds, and a correspondingly large sequential running time. Our goal will now be to reduce the width to a constant.

In order to achieve this, consider the following modified polyhedral constraints, that we term $P(D, q)$. Here, $q \geq 1$ will be a small integer. We will denote the corresponding feasibility problem as $\text{DUAL}(D, q)$.

$$\sum_{e \text{ incident on } v} \alpha_{ev} \leq D \ \forall v \in V$$
$$\alpha_{ev} \leq q \ \forall e \in E, v \in V$$
$$\alpha_{ev} \geq 0 \ \forall e \in E, v \in V$$

Note that the second constraint in $P(D, q)$ is new, and it does not change the feasibility of $\text{DUAL}(D)$, since if the original system is feasible, it is also feasible with $\alpha_{ev} \leq 1$ for all e, v. In other words, $\text{DUAL}(D)$ is feasible iff $\text{DUAL}(D, 1)$ is feasible. We will set q to be a small constant that we decide later.

Lemma 2. *The width ρ of $\text{DUAL}(D, q)$ as written above is at most $2q$.*

Proof. For any $\alpha \in P(D, q)$, we have $\alpha_{eu} \leq q$. Therefore $\alpha_{eu} + \alpha_{ev} \leq 2q$, which implies a width of $2q$.

In order to apply the multiplicative weight update method as described in Appendix A, we need to compute $\text{ORACLE}(\mathbf{y})$. This involves solving the following problem for given $\mathbf{y}$:

$$C(\mathbf{y}, D, q) = \max_{\alpha \in P(D, q)} \sum_{v} \sum_{e \text{ incident on } v} y_e \alpha_{ev}$$

Lemma 3. $\text{ORACLE}(\mathbf{y})$ *can be computed in $O(m)$ time.*

Proof. For any $\mathbf{y}$, for each v, the optimal solution $C(\mathbf{y}, D, q)$ sets α_{ev} as follows. Let $r = \lfloor D/q \rfloor$ and let $s = D - rq$. Then $\alpha_{ev} = q$ for the r largest y_e incident on v, and $\alpha_{ev} = s$ for the e with the $(r + 1)^{st}$ largest y_e. This involves finding the rth and $(r + 1)^{st}$ largest y_e for each vertex, which can be done in linear time. This is followed by a linear time computation to set the values.

Using the above two lemmas, the following theorem is immediate from Theorem 5.

Theorem 1. *For any integer D and constants q and $\epsilon \in [0,1]$, in time $O\left(m\frac{\log m}{\epsilon^2}\right)$, the multiplicative weight algorithm either returns that $\mathrm{DUAL}(D,q)$ is infeasible, or finds α so that for all $e = (u,v) \in E$: $\alpha_{eu} + \alpha_{ev} \geq 1 - \epsilon$.*

2.3 Binary Search for D^*

We showed earlier how we can apply the multiplicative weight update algorithm to decide if $\mathrm{DUAL}(D,q)$ is feasible. Let $k(\epsilon)$ be the number of phases needed to compute a $(1+\epsilon)$ approximation to D^*, and $k'(\epsilon)$ the number of phases needed to compute a $(1+\epsilon)$ approximation given a $(1+2\epsilon)$ approximation, where $\epsilon < 1/2$. From Theorem 5, it follows (since q is a constant) that $k'(\epsilon) = O\left(\frac{\log n}{\epsilon^2}\right)$. Since $k(\epsilon) \leq k'(\epsilon) + k(2\epsilon)$, this gives $k(\epsilon) = O\left(\frac{\log n}{\epsilon^2}\right)$. Since ϵ decreased by a factor of 2 in each recursive step above, the above recurrence can be thought of as a binary search.

Let $(\alpha, \mathbf{y})$ denote the final solution corresponding to running the multiplicative weight procedure on $\mathrm{DUAL}(\tilde{D}, q)$. Here the approximately optimal dual solution $\mathbf{y}$ is found as in Theorem 5.

Theorem 2. *For $0 \leq \epsilon \leq 1/3$ and any constant $q \geq 1$, the value $\tilde{D}$ and the final solution $(\alpha, \mathbf{y})$ satisfy:*

1. $D^*(1 - \epsilon) \leq \tilde{D} \leq D^*(1 + \epsilon)$.
2. $\sum_e y_e \geq (1 - 3\epsilon)C(\mathbf{y}, \tilde{D}, q)$.

Proof. Suppose $\tilde{D} < D^*(1-\epsilon)$. Since the multiplicative weight procedure returns an ϵ-optimal solution, we can scale up α by $1/(1-\epsilon)$ so that these values are feasible for $\mathrm{DUAL}(D, 2q)$ for $D = \tilde{D}/(1-\epsilon) < D^*$, and hence feasible for $\mathrm{DUAL}(D)$. This violates the optimality of D^* as the smallest D for which $\mathrm{DUAL}(D)$ is feasible. On the other hand, $\mathrm{DUAL}(D, q)$ is feasible for any $D \in [D^*, D^*(1 + \epsilon)]$, which means the multiplicative weight algorithm cannot declare in-feasibility for any D falling within this range. Therefore $D^*(1 - \epsilon) \leq \tilde{D} \leq D^*(1 + \epsilon)$.

Recall from the discussion preceding Theorem 5 that

$$\lambda^* = \max\{\lambda \ \ s.t. \ \ \alpha_{eu} + \alpha_{ev} \geq \lambda \ \ \forall e = (u,v) \text{ is feasible for } \alpha \in P(D,q)\}$$

For $D \leq D^*(1+\epsilon)$, we have $\lambda^* \leq 1+\epsilon$, else by scaling α we can show $\mathrm{DUAL}(D,q)$ is feasible for $D < D^*$, which is not possible. Therefore, Theorem 5 implies for $0 \leq \epsilon \leq 1/3$:

$$\sum_e y_e \geq (1 - \epsilon)^2 C(y, \tilde{D}, q) \geq (1 - 3\epsilon)C(y, \tilde{D}, q)$$

2.4   Rounding Step: Recovering the Densest Subgraph

Using Theorem 2, we have a value $\tilde{D} \in [(1 - \epsilon)D^*, (1 + \epsilon)D^*]$ along with dual variables $\mathbf{y}$ that satisfies $\sum_e y_e \geq (1 - 3\epsilon)C(\mathbf{y}, \tilde{D}, q)$. We will now use these variables to recover an approximately optimal densest subgraph. We first discuss why this is not straightforward.

Technical Hurdle. The problem $\text{DUAL}(\tilde{D}, q)$ is different from the problem $\text{DUAL}(\tilde{D})$ in that the corresponding primal problems are different. The primal feasibility problem corresponding to $\text{DUAL}(\tilde{D}, q)$ is the following:

$$\text{Find } \mathbf{y}, \mathbf{x}, \mathbf{z} \geq 0 \text{ s.t.} \qquad \frac{\sum_e y_e}{\sum_{e,v}(\tilde{D}x_v + qz_{ev})} \geq 1$$

$$\text{where} \qquad y_e \leq \min\left(x_u + z_{eu}, x_v + z_{ev}\right) \quad \forall e = (u, v) \in E$$

The primal feasibility problem corresponding to $\text{DUAL}(\tilde{D})$ does not have variables $\mathbf{z}$. These problems are equivalent from the perspective of exact feasibility since the *exact* optimal primal solution of $\text{DUAL}(\tilde{D}, q)$ will indeed set $\mathbf{z} = 0$, and the resulting $\mathbf{x}, \mathbf{y}$ are precisely the vertex and edge variables in the LP formulation we began from. The catch is the following: *An ϵ-approximate solution using $\text{DUAL}(\tilde{D}, q)$ need not yield an ϵ-approximate solution using $\text{DUAL}(\tilde{D})$.* The reason is that an approximately optimal solution to $\text{DUAL}(\tilde{D}, q)$ might have large $\mathbf{z}$, so that the resulting $\mathbf{y}, \mathbf{x}$ variables have no easy interpretation.

Despite this difficulty, we show that for $q = 2$, we can *round* an ϵ-approximate solution to $\text{DUAL}(\tilde{D}, q)$ into an ϵ-approximate solution to $\text{DUAL}(\tilde{D})$, and hence recover the approximate densest subgraph. We note that the primal problem itself is a fractional densest subgraph that must be further converted (or rounded) into an integer solution. We fold both the rounding steps into one in the proof below, *noting that even recovering the fractional densest subgraph would need our new rounding method.*

First recall that $C(\mathbf{y}, \tilde{D}, q)$ is computed as follows: Let $\tilde{r} = \lfloor \tilde{D}/q \rfloor$, and $\tilde{s} = \tilde{D} - q\tilde{r}$. For simplicity in the proof below, we assume $\tilde{s} > 0$. For any vertex v, we sort the y_e values incident on v in decreasing order, and denote these $y_1(v) \geq y_2(v) \geq \cdots \geq y_n(v)$. Then:

$$C(\mathbf{y}, \tilde{D}, q) = \sum_v \left(\sum_{k=1}^{\tilde{r}} qy_k(v) + \tilde{s}y_{\tilde{r}+1}(v) \right)$$

The important point is that this is a linear function of $\mathbf{y}$.

Step 1: Discretization. This step is mainly to improve efficiency. Let $Y = \max_e y_e$. Scale up or down the y_e values so that $Y = 1$. Consider all edges e with $y_e \leq \epsilon/m^2$. The contribution of these edges to the summation $\sum_e y_e$ and to $C(y, \tilde{D}, q)$ is at most ϵ/m. We set all these $y_e = 0$. Since we originally had $\sum_e y_e \geq (1-3\epsilon)C(\mathbf{y}, \tilde{D}, q)$, the new vector $\mathbf{y}$ satisfies: $\sum_e y_e \geq (1-4\epsilon)C(\mathbf{y}, \tilde{D}, q)$.

Now round each y_e down to the nearest power of $(1+\epsilon)$. This does not change any individual y_e by more than a factor of $(1 + \epsilon)$. Therefore, the resulting $\mathbf{y}$

satisfies: $\sum_e y_e \geq (1 - 6\epsilon)C(\mathbf{y}, \tilde{D}, q)$. At this point, note that there are only $O\left(\frac{\log m}{\epsilon}\right)$ distinct values of y_e.

Step 2: Line Sweep. Fix any $\gamma \geq 0$. Let $I(z) = 1$ if $z \geq \gamma$. Consider the process that includes edge e if $y_e \geq \gamma$. Let $G(\gamma)$ denote the subgraph induced by these edges; let $E(\gamma)$ denote the set of induced edges; $V(\gamma)$ denote the set of induced vertices; and let $d_v(\gamma)$ denote the degree of v in $G(\gamma)$. Note that $d_v(\gamma) = \sum_{e \in N(v)} I(y_e)$, and $|E(\gamma)| = \sum_e I(y_e)$. Furthermore, let:

$$H_v(\gamma) = \sum_v \left(\sum_{k=1}^{\tilde{r}} I(y_k(v)) + \frac{\tilde{s}}{q}I(y_{\tilde{r}+1}(v))\right)$$

Lemma 4. *There exists γ such that $G(\gamma)$ is non-empty, and $|E(\gamma)| \geq q(1 - 6\epsilon) \sum_v H_v(\gamma)$. Furthermore, this value of γ can be computed in $O\left(m \frac{\log m}{\epsilon}\right)$ time.*

Proof. We note that:

$$\sum_e y_e = \int_{\gamma=0}^1 |E(\gamma)| d\gamma$$

$$C(y, \tilde{D}, q) = q \int_{\gamma=0}^1 \sum_v H_v(\gamma) d\gamma$$

Since $\sum_e y_e \geq (1 - 6\epsilon)C(y, \tilde{D}, q)$, this implies the existence of a γ that satisfies the condition of the lemma. There are only $O\left(\frac{\log m}{\epsilon}\right)$ distinct values of γ, and computing $|E(\gamma)|$, $\sum_v \min(\tilde{D}/q, d_v(\gamma))$ takes $O(m)$ time for any γ.

Start with the value of γ that satisfies $|E(\gamma)| \geq q(1 - 6\epsilon) \sum_v H_v(\gamma)$. Let V_1 denote the set of vertices such that for $v \in V_1$, we have $y_{\tilde{r}+1}(v) \geq \gamma$. For these vertices, $H_v(\gamma) = \sum_{k=1}^{\tilde{r}} 1 + \tilde{s}/q = \tilde{D}/q$. Let V_2 denote the remaining vertices; for these we have $d_v(\gamma) = H_v(\gamma)$. Therefore, we have

$$\sum_v H_v(\gamma) = \tilde{D}/q \times |V_1| + \sum_{v \in V_2} d_v(\gamma)$$

Suppose we delete all vertices in V_2 simultaneously. Let $G(V_1, E_1)$ denote the subgraph induced on V_1. Then:

$$|E_1| \geq |E(\gamma)| - \sum_{v \in V_2} d_v(\gamma) \geq q(1 - 6\epsilon)\left(\tilde{D}/q \times |V_1| + \sum_{v \in V_2} d_v(\gamma)\right) - \sum_{v \in V_2} d_v(\gamma)$$

Therefore, $|E_1| \geq (1 - 6\epsilon)\tilde{D}|V_1|$ for $q \geq 2$ and $\epsilon < 1/12$.

The final technicality is to show that $G(V_1, E_1)$ is non-empty. There are now two cases: (a) If $\sum_{v \in V_2} d_v(\gamma) = 0$, then $|E_1| \geq |E(\gamma)| > 0$, so that $G(V_1, E_1)$ is non-empty. (b) Otherwise, the final inequality is strict and we again have $|E_1| > 0$. This implies $G(V_1, E_1)$ is always non-empty. The density of $G(V_1, E_1)$ is at least $\tilde{D}(1 - 6\epsilon) \geq D^*(1 - 7\epsilon)$, and we finally have the following theorem.

Theorem 3. *For $\epsilon \in (0, 1/12)$, a subgraph of density $D^*(1 - \epsilon)$ can be computed in $O\left(m \frac{\log m}{\epsilon^2}\right)$ time.*

The key point in the above proof is that the very final inequality crucially needs $q > 1 + 6\epsilon$; indeed for smaller values of q, there are examples where an approximately optimal solution to $\text{DUAL}(\tilde{D}, q)$ does not imply an approximately optimal densest subgraph in any natural way.

2.5   Summary of the Algorithm

Before presenting the MAPREDUCE implementation details, we summarize the algorithm as follows:

- Define $\text{DUAL}(D, q)$ and $\text{ORACLE}(\mathbf{y}) = C(\mathbf{y}, D, q)$ for $q = 2$.
- Decide feasibility of $\text{DUAL}(D, q)$ using multiplicative weight method; wrap this in a discretized binary search to find the smallest $D = \tilde{D}$ for which the problem is approximately feasible.
- The output of the previous step is a value $\tilde{D}$ and dual vector $\mathbf{y}$ such that $\sum_e y_e \geq (1 - 3\epsilon)C(\mathbf{y}, \tilde{D}, q)$.
- Discretize y_e and throw away values at most ϵ/m^2 times the largest value.
- Perform a line sweep to find γ for which the subgraph induced by edges e with $y_e \geq \gamma$ satisfies the condition in Lemma 4.
- Remove vertices with degree at most $\lfloor \tilde{D}/q \rfloor$ from this subgraph, and output the remaining subgraph.

2.6   Number of MAPREDUCE Phases

We now show how to implement the above algorithm in $O\left(\frac{\log m}{\epsilon^2}\right)$ MAPREDUCE phases, where each phase operates on $O(m)$ total (key; value) pairs.

Oracle $C(\mathbf{y}, D, q)$ Computation. The mappers take as input $(e; \alpha_{eu}, \alpha_{ev}, y_e^{old})$, and produce $(u; y_e^{new})$ and $(v; y_e^{new})$. These are shuffled using the vertex as the key. The reducer for vertex v needs to compute the $\lceil D/q \rceil^{th}$ largest y_e and set α_{ev} for edges e with larger y_e. This takes linear time in d_{max}, the maximum degree; this determines the reducer size. The reducers output $\{(e; \alpha_{ev}, y_e)\}$ and $(v; S_v)$, where S_v is the contribution of v to $C(\mathbf{y}, D, q)$. The next phase does a summation of the S_v to compute the value of the oracle; the next shuffle phase shuffles the $(e; \alpha_{ev}, y_e)$ using the edge as the key; and the next reduce phase produces $(e; \alpha_{eu}, \alpha_{ev}, y_e)$. Therefore, the oracle can be implemented in two MAPREDUCE phases.

Binary Search. As mentioned before, this takes $k(\epsilon) = O\left(\frac{\log m}{\epsilon^2}\right)$ phases of the oracle computation.

Rounding Steps. The scaling requires estimating the maximum of the y_e, which requires a single phase. The scaling and discretization can be done in the next Map phase. The various choices of γ can be tried in sequence, one in each phase of MapReduce, giving $O\left(\frac{\log m}{\epsilon}\right)$ phases. For each γ, the subgraph computed involves filtering the edges by their y_e and throwing away small degree vertices; this takes $O(m)$ sequential complexity and $O(d_{\max})$ reducer size. One final MapReduce phase can pick the best density from the various values of γ. We therefore have the following theorem:

Theorem 4. *For $\epsilon \in (0, 1/12)$, a subgraph of density $D^*(1-\epsilon)$ can be computed with $O(d_{\max})$ reducer size, $O(m)$ sequential complexity per phase, and $O\left(\frac{\log m}{\epsilon^2}\right)$ phases.*

Note that restricting ϵ to be less than $1/12$ poses no problem in the binary search since we can always start with an $O(1)$ approximation using [5]. We finally note that the same technique applies to the *weighted* version of the problem.

A   The Multiplicative Weights Update Framework

We next present a simple algorithm for deciding feasibility of covering linear programs due to Young [17]; we use the exposition and bounds due to Arora-Hazan-Kale [3]. We first define the generic covering problem.

COVERING: $\exists? \mathbf{x} \in P$ such that $A\mathbf{x} \geq \mathbf{1}$, where A is an $r \times s$ matrix and P is a convex set in $\mathbf{R}^s$ such that $A\mathbf{x} \geq \mathbf{0}$ for all $\mathbf{x} \in P$.

The running time of the algorithm is quantified in terms of the WIDTH defined as:

$$\rho = \max_i \max_{\mathbf{x} \in P} \mathbf{a}_i \mathbf{x}$$

The algorithm assumes an efficient oracle that solves the *Lagrangian* of the constraints $A\mathbf{x} \geq \mathbf{1}$. Given dual multipliers y_i associated with the constraint $\mathbf{a}_i\mathbf{x} \geq 1$, the oracle takes a linear combination of the constraints, and maximizes the LHS with respect to $\mathbf{x}$.

ORACLE($\mathbf{y}$): Given an r-dimensional dual vector $\mathbf{y} \geq 0$, solve $C(\mathbf{y}) = \max\{\mathbf{y}^t A\mathbf{z} : \mathbf{z} \in P\}$.

By duality theory, and as explained in [3], it is easy to see that if there exists $\mathbf{y} \geq \mathbf{0}$, $C(\mathbf{y}) < ||\mathbf{y}||_1$ (where $||\mathbf{y}||_1 = \mathbf{y}^t\mathbf{1} = \sum_{k=1}^r y_r$ is the l_1 norm of vector $\mathbf{y}$), then $A\mathbf{x} < 1$ for all $\mathbf{x} \in P$, and hence COVERING is infeasible. The multiplicative weight update procedure described in Fig. 1 iteratively updates the vector $\mathbf{y}$ so that it either correctly declares in-feasibility of the program or finds an $\mathbf{x} \in P$ with $A\mathbf{x} \geq (1-\epsilon)\mathbf{1}$.

The above procedure not only provides a guarantee on the final value $\mathbf{x}^*$, but also yields a guarantee on the dual variables $\mathbf{y}$ that it computes. Define an optimization version of COVERING as follows:

$$\lambda^* = \max\{\lambda \; : \; A\mathbf{x} \geq \lambda\mathbf{1} \text{ and } \mathbf{x} \in P\}$$

MULTIPLICATIVE WEIGHT UPDATE ALGORITHM
Let $T \leftarrow \frac{4\rho \log r}{\epsilon^2}$; $\mathbf{y}_1 = \mathbf{1}$
For $k = 1$ **to** T **do:**
  Find $\mathbf{x}_k$ using ORACLE$(\mathbf{y}_k)$.
  If $C(\mathbf{y}_k) < ||\mathbf{y}_k||_1$, then declare **infeasible** and stop.
  Update $y_{ik+1} \leftarrow y_{ik}\left(1 - \epsilon\frac{\mathbf{a}_i \mathbf{x}_k}{\rho}\right)$ for $i = 1, 2, \ldots, r$.
Return $\mathbf{x}^* = (\sum_k \mathbf{x}_k)/T$.

Fig. 1. The Multiplicative Weight Update Algorithm for Covering Linear Programs

The problem COVERING is equivalent to deciding $\lambda^* \geq 1$. By the definition of λ^*, we have that for any $\mathbf{y} \geq 0$, $\lambda^*||\mathbf{y}||_1 \leq C(\mathbf{y})$; the multiplicative weight method makes this inequality approximately tight as well if it runs for T steps. The next theorem is implicit in the analysis of [3], and is explicitly presented in [8,17].

Theorem 5. *The multiplicative weight procedure either correctly outputs $A\mathbf{x} \geq \mathbf{1}$ is infeasible for $x \in P$, or finds a solution $\mathbf{x}^* \in P$ such that $A\mathbf{x}^* \geq (1 - \epsilon)\mathbf{1}$. Furthermore, in the latter case, there exists an iteration k such that[3]:*

$$\lambda^* \times ||\mathbf{y}_k||_1 \geq (1 - \epsilon)C(\mathbf{y}_k)$$

B Densest Subgraph in Directed Graphs

Let $G(V, E)$ denote a directed graph. The densest subgraph problem asks to find two subsets $S, T \subseteq V$ (not necessarily disjoint) such that if $E(S, T)$ denotes the set of edges (u, v) where $u \in S$ and $v \in T$, then the density $D(S, T) = \frac{|E(S,T)|}{\sqrt{|S||T|}}$ is maximized. Let OPT denote the optimal density. We present an algorithm with running time $O\left(m\frac{\log^2 m}{\epsilon^3}\right)$ that outputs a subgraph of density $(1 - \epsilon)OPT$. We will only outline the portions different from the undirected case.

B.1 Parametric LP Formulation

We start with a modification of an ingenious parametrized linear program of Charikar [6]. Consider the following linear program PRIMAL(z). There is a variable y_e for each edge $e \in E$, and variables s_v, t_v for each vertex $v \in V$. The value z is a parameter that we will search over.

$$\text{Maximize} \sum_e y_e$$

---

[3] Young [17] shows that if $A\mathbf{x} \geq \lambda\mathbf{1}$ for the final solution, then the average over k of $C(\mathbf{y}_k)/||\mathbf{y}_k||_1$ converges to at most $\lambda/(1 - \epsilon) \leq \lambda^*/(1 - \epsilon)$. This implies one of these values is at most $\lambda^*/(1 - \epsilon)$, implying the claim. The exact same guarantee is also explicitly obtained by Plotkin-Shmoys-Tardos [14], though their update procedure is somewhat different and much more involved.

$$y_e \le s_u \ \forall e = (u, v) \in E$$
$$y_e \le t_v \ \forall e = (u, v) \in E$$
$$\sum_v \left(z s_v + \tfrac{1}{z} t_v \right) \le 2$$
$$y_e, s_v, t_v \ge 0 \quad \forall e, v$$

The difference with the LP in [6] is that our penultimate constraint is obtained by taking a linear combination of two constraints in their LP. This leads to a nicer dual formulation where the dual objective can be directly interpreted as the density.

Lemma 5 ([6]). $OPT \le \max_z \text{PRIMAL}(z)$.

Proof. Let $S, T \subseteq V$ denote the densest subgraph. Set $z = \sqrt{|T|/|S|}$. Set $y_e = 1/\sqrt{|S||T|}$ for all e within the subgraph; set $s_u = 1/\sqrt{|S||T|}$ for all $u \in S$, and set $t_v = 1/\sqrt{|S||T|}$ for all $v \in T$. It is easy to check that this is a feasible solution with objective value exactly equal to the density of this subgraph.

Let $D^* = \max_z \text{PRIMAL}(z)$. Consider the following dual linear program DUAL(D, z):

$$\alpha_{eu} + \alpha_{ev} \ge 1 \qquad \forall e = (u, v) \in E$$
$$\sum_{e|e=(v,w)} \alpha_{ev} \le Dz/2 \quad \forall v \in V$$
$$\sum_{e|e=(u,v)} \alpha_{ev} \le D/(2z) \ \forall v \in V$$
$$\alpha_{ev} \ge 0 \qquad \forall e, v$$

Using strong duality, we immediately have:

Lemma 6. $D^* = \max_z \min\{D \mid \text{DUAL}(D, z) \text{ is feasible}\}$.

B.2 Covering Program and Width Modulation

In order to decide if DUAL(D, z) is feasible, we use the multiplicative weight framework to decide the feasibility of the set of constraints:

$$\alpha_{eu} + \alpha_{ev} \ge 1 \qquad \forall e = (u, v) \in E$$

subject to the polyhedral constraints $P(D, z)$:

$$\sum_{e|e=(v,w)} \alpha_{ev} \le Dz/2 \quad \forall v \in V$$
$$\sum_{e|e=(u,v)} \alpha_{ev} \le D/(2z) \ \forall v \in V$$
$$\alpha_{ev} \le 2 \qquad \forall e, v$$
$$\alpha_{ev} \ge 0 \qquad \forall e, v$$

As before, the constraints $\alpha_{ev} \le 2$ have been added to reduce the width of the program. Note that the width is at most 4. The ORACLE($\mathbf{y}$) problem assigns dual vector $\mathbf{y}$ to the set of constraints $\alpha_{eu} + \alpha_{ev} \ge 1$ and solves:

$$C(\mathbf{y}, D, z) = \max_{\alpha \in P(D,z)} \sum_v \sum_{e \text{ incident on } v} y_e \alpha_{ev}$$

This can be solved in $O(m)$ time: For each vertex v, we only need to find the top $Dz/4$ values y_e for edges leaving v, and the top $D/(4z)$ values y_e for edges entering v; we set $\alpha_{ev} = 2$ for the corresponding edges. We assume for simplicity of exposition that $Dz/4$ and $D/(4z)$ are integers; the proof extends to the general case with minor modification.

Lemma 7. *For any D, z and $\epsilon \in [0, 1]$, in time $O\left(m\frac{\log m}{\epsilon^2}\right)$, the multiplicative weight update algorithm either returns that $\text{DUAL}(D, z)$ is infeasible, or finds $\alpha \in P(D, z)$ such that $\alpha_{eu} + \alpha_{ev} \geq 1 - \epsilon$ for all $e = (u, v) \in E$.*

B.3   Parametric Search

We apply the multiplicative weight update algorithm within the following parametric search procedure. Discretize z and D in powers of $(1 + \delta)$. For each discretized z, perform a binary search to find the smallest discretized D (call this $\tilde{D}(z)$) for which the multiplicative weight algorithm returns a feasible (δ-optimal) solution. Find that z which maximizes $\tilde{D}(z)$; denote this value of z as $\tilde{z}$ and this value of $\tilde{D}(z)$ as $\tilde{D}$. Since the density is upper bounded by m, and since we can assume D/z and Dz lie in $[1, 2n]$, the number of parameters we try is $O\left(\frac{\log m}{\delta}\right)$ ignoring lower order terms. This increases the running time by the corresponding factor compared to the undirected case.

Let $(\alpha, \mathbf{y})$ denote the final solution found by the above procedure, where the dual vector $\mathbf{y}$ is found as in Theorem 5. The following theorem is proved analogously to Theorem 2 by choosing δ to be a sufficiently small constant fraction of ϵ. (The only additional observation we need is that modifying z by a factor of $(1 + \delta)$ changes $\tilde{D}(z)$ by at most that factor.)

Theorem 6. *For $0 \leq \epsilon \leq 1/3$, the values $\tilde{D}, \tilde{z}$ and the final solution $(\alpha, \mathbf{y})$ satisfy:*

1. $D^*(1 - \epsilon) \leq \tilde{D} \leq D^*(1 + \epsilon)$.
2. $\sum_e y_e \geq (1 - 3\epsilon)C(y, \tilde{D}, \tilde{z})$

B.4   Rounding Step: Recovering the Densest Subgraph

Using Theorem 2, we have a value $\tilde{D} \in [(1 - \epsilon)D^*, (1 + \epsilon)D^*]$ along with a value $\tilde{z}$ and dual variables $\mathbf{y}$ that satisfy $\sum_e y_e \geq (1 - 3\epsilon)C(\mathbf{y}, \tilde{D}, \tilde{z})$. We will now use these variables to recover an approximately optimal densest subgraph.

Step 1: Discretization. First note that $C(\mathbf{y}, \tilde{D}, \tilde{z})$ is computed as follows: For each $v \in V$, sum up the largest at most $\tilde{D}\tilde{z}/4$ y_e for $e = (v, w)$, and the largest $\tilde{D}/(4\tilde{z})$ y_e for $e = (w, v)$, and double this value. Let $Y = \max_e y_e$. Scale up or down the y_e values so that $Y = 1$. We eliminate all edges with $y_e \leq \epsilon/m^2$ and round each y_e down to the nearest power of $(1 + \epsilon)$. As shown before, the resulting $\mathbf{y}$ satisfies:

$$\sum_e y_e \geq (1 - 6\epsilon)C(\mathbf{y}, \tilde{D}, \tilde{z})$$

At this point, note that there are only $O\left(\frac{\log m}{\epsilon}\right)$ distinct values of y_e.

Step 2: Line Sweep. Let $G(\gamma)$ denote the subgraph induced by edges with $y_e \geq \gamma$. Let $E(\gamma)$ denote the set of induced edges, $S(\gamma)$ denote the set of induced source vertices, and $T(\gamma)$ denote the set of induced destination vertices. Let $d_v^S(\gamma)$ and $d_v^T(\gamma)$ denote the out-degree and in-degree respectively of v in $G(\gamma)$.

Lemma 8. *There exists γ such that*

$$|E(\gamma)| \geq 2(1 - 6\epsilon) \sum_v \left(\min\left(\tilde{D}\tilde{z}/4, d_v^S(\gamma) \right) + \min\left(\tilde{D}/(4\tilde{z}), d_v^T(\gamma) \right) \right)$$

Furthermore, this value of γ can be computed in $O\left(m \frac{\log m}{\epsilon} \right)$ time.

Proof. We note that:

$$\sum_e y_e = \int_{\gamma=0}^1 |E(\gamma)| d\gamma$$

and

$$C(\mathbf{y}, \tilde{D}, \tilde{z}) = 2 \int_{\gamma=0}^1 \sum_v \left(\min\left(\tilde{D}\tilde{z}/4, d_v^S(\gamma) \right) + \min\left(\tilde{D}/(4\tilde{z}), d_v^T(\gamma) \right) \right) d\gamma$$

Since $\sum_e y_e \geq (1 - 6\epsilon)C(\mathbf{y}, \tilde{D}, \tilde{z})$, this implies the existence of a γ that satisfies the condition of the lemma. There are only $O\left(\frac{\log m}{\epsilon} \right)$ distinct values of γ, and computing $|E(\gamma)|$ and the degrees of the vertices takes $O(m)$ time for any γ.

Start with the value of γ that satisfies the condition of the previous lemma. Let S_1 denote the set of vertices with $d_v^S(\gamma) \geq \tilde{D}\tilde{z}/4$ and T_1 denote the set of vertices with $d^T(\gamma) \geq \tilde{D}/(4\tilde{z})$. We have:

$$\sum_v \left(\min\left(\frac{\tilde{D}\tilde{z}}{4}, d_v^S(\gamma) \right) + \min\left(\frac{\tilde{D}}{4\tilde{z}}, d_v^T(\gamma) \right) \right) = \frac{\tilde{D}}{4}\left(\tilde{z}|S_1| + \frac{|T_1|}{\tilde{z}} \right) + \sum_{v \in V \setminus S_1} d_v^S(\gamma) + \sum_{v \in V \setminus T_1} d_v^T(\gamma)$$

Consider the sets (S_1, T_1) and the edge set $E(S_1, T_1)$ that goes from S_1 to T_1. Since this edge set is obtained by deleting edges out of $V \setminus S_1$ and edges into $V \setminus T_1$, the above conditions imply:

$$|E(S_1, T_1)| \geq |E(\gamma)| - \left(\sum_{v \in V \setminus S_1} d_v^S(\gamma) + \sum_{v \in V \setminus T_1} d_v^T(\gamma) \right)$$

$$\geq 2(1 - 6\epsilon) \left(\frac{\tilde{D}}{4}\left(\tilde{z}|S_1| + \frac{|T_1|}{\tilde{z}} \right) + \sum_{v \in V \setminus S_1} d_v^S(\gamma) + \sum_{v \in V \setminus T_1} d_v^T(\gamma) \right)$$

$$- \left(\sum_{v \in V \setminus S_1} d_v^S(\gamma) + \sum_{v \in V \setminus T_1} d_v^T(\gamma) \right)$$

$$\geq (1 - 6\epsilon)\frac{\tilde{D}}{2}\left(\tilde{z}|S_1| + |T_1|/\tilde{z} \right)$$

for $\epsilon \leq 1/12$. Now observe that for any $\tilde{z}$, we have $\tilde{z}|S_1| + |T_1|/\tilde{z} \geq 2\sqrt{|S||T|}$. Therefore:

$$\text{Density of } (S_1, T_1) = \frac{|E(S_1, T_1)|}{\sqrt{|S||T|}} \geq (1 - 6\epsilon)\tilde{D} \geq (1 - 7\epsilon)D^*$$

Theorem 7. *For any $\epsilon \in (0, 1/12)$, a directed subgraph with density $D^*(1 - \epsilon)$ can be computed in time $O\left(m\frac{\log^2 m}{\epsilon^3}\right)$.*

MAPREDUCE *Implementation.* The details of the ORACLE computation and the rounding step are the same as the undirected case. The parametric search can be done in parallel for all values of D, z; the sequential complexity per phase is now $O\left(m\frac{\log^2 m}{\epsilon^2}\right)$ since the parametric search is over both D and z instead of just over D. Therefore, the overall implementation still takes $O\left(\frac{\log m}{\epsilon^2}\right)$ phases with $O\left(m\frac{\log^2 m}{\epsilon^2}\right)$ sequential complexity per phase, and $O(d_{\max})$ reducer size, where $d_{\max}$ is the maximum degree.

C   Fractional Matchings in Bipartite Graphs

In this section, we show the broader applicability of width modulation by presenting an FPTAS for fractional maximum size matchings in bipartite graphs; incidentally, we also obtain a slight decrease in the number of phases. As motivation, consider the following Adword allocation problem [13]: There are n_1 advertisers, where advertiser u has demand d_u; there are n_2 ad slots, where slot v has supply s_v; furthermore there is a bipartite graph specifying which advertiser is interested in which ad slots. The goal is to find a (possibly fractional) allocation that maximizes the demand satisfied.

We present an efficient algorithm for the case where all supplies and demands are 1 (the maximum fractional matching case); this algorithm extends with minor modification to the general case. On a graph with n vertices and m edges, our algorithm takes $O\left(m\frac{\log n}{\epsilon^3}\right)$ running time, and can be implemented in $O\left(\frac{\log^2 n}{\epsilon^3 \log d_{\max}}\right)$ phases of MAPREDUCE with reducer size $O(d_{\max})$, the maximum degree, and sequential complexity $\tilde{O}(m)$ per phase. The number of phases improves by a $\log d_{\max}$ factor that which can be achieved by the algorithm of Ahn-Guha [2], which was designed for the semi-streaming model.

Formally, we are given a bipartite graph $G(V_1, V_2, E)$ on n vertices. Let K^* denote the size of the optimal fractional matching. For vertex v, let $N(v)$ denote the set of edges incident on v. The maximum fractional matching is the solution to the following linear program.

$$\text{Maximize } \sum_e y_e$$

$$\sum_{e \in N(u)} y_e \leq 1 \; \forall u \in V_1$$
$$\sum_{e \in N(v)} y_e \leq 1 \; \forall v \in V_2$$
$$y_e \geq 0 \; \forall e \in E$$

As before, we write the following dual program $\text{DUAL}(K)$.

$$x_u + x_v \geq 1 \;\; \forall e = (u, v) \in E$$
$$\sum_{v \in V_1 \cup V_2} x_v \leq K$$

It is well-known that the dual is fractional VERTEX COVER, where variable x_v captures whether vertex v is in the cover. By strong duality, $\text{DUAL}(K)$ is feasible iff $K \geq K^*$.

C.1 Covering Program, Width Modulation, and Binary Search

We will now check the feasibility of the constraints:

$$x_u + x_v \geq 1 \qquad \forall e = (u, v) \in E$$

subject to the polyhedral constraints $P(K)$.

$$\sum_v x_v \leq K$$
$$x_v \leq 1/\epsilon \; \forall v \in V_1 \cup V_2$$

As before, the second set of constraints is added to reduce the width of the program. Without these constraints, the width can be as large as $2K$. With these constraints, it becomes at most $2/\epsilon$. For simplicity of exposition, we assume from now that ϵK is an integer; the proof only needs a minor modification if this assumption is removed.

The problem $\text{ORACLE}(\mathbf{y})$ assigns dual variables $\mathbf{y}$ to the set of constraints $x_u + x_v \geq 1$, and computes:

$$C(\mathbf{y}, K) = \max_{\mathbf{x} \in P(K)} \sum_{v \in V_1 \cup V_2} x_v \left(\sum_{e \in N(v)} y_e \right)$$

This involves the following $O(m)$ time computations: (1) Compute $S_v(\mathbf{y}) = \sum_{e \in N(v)} y_e$ for each vertex v; and (2) Compute the top ϵK values $S_v(\mathbf{y})$; sum these; and multiply this result by $1/\epsilon$.

The following theorem is now analogous to Theorem 1

Theorem 8. *For any integer K and $\epsilon \in [0, 1]$, in time $O\left(m\frac{\log n}{\epsilon^3}\right)$, the multiplicative weight algorithm either returns that $\text{DUAL}(K)$ is infeasible, or finds $\mathbf{x} \in P(K)$ such that $x_u + x_v \geq 1 - \epsilon$ for all $e = (u, v) \in E$.*

As before, we apply the multiplicative weight algorithm in the following *Binary Search* procedure: Round the values K in powers of $(1 + \epsilon)$. Perform binary search over these discretized values to find the smallest K (call this $\tilde{K}$)

for which the multiplicative weight algorithm returns a feasible solution. Let $(\mathbf{x}, \mathbf{y})$ denote the final solution found by the above procedure, where the dual solution $\mathbf{y}$ is found as in Theorem 5. The following theorem is now analogous to Theorem 2.

Theorem 9. *For $0 \leq \epsilon \leq 1/3$, the value $\tilde{K}$ and the final solution $(\mathbf{x}, \mathbf{y})$ satisfy:* *(1) $K^*(1 - \epsilon) \leq \tilde{K} \leq K^*(1 + \epsilon)$; and (2) $\sum_e y_e \geq (1 - 3\epsilon)C(y, \tilde{K})$.*

C.2   Rounding Step: Recovering the Fractional Matching

We start with the $\tilde{K} \in [K^*(1-\epsilon), K^*(1+\epsilon)]$ and $\mathbf{y}$ computed above. Recall that $S_v(\mathbf{y}) = \sum_{e \in N(v)} y_e$. Therefore, $\sum_{e \in E} y_e = \frac{1}{2}\sum_v S_v(\mathbf{y})$. Let $\kappa = \epsilon \tilde{K}$. Suppose the vertices are sorted in decreasing order of $S_v(\mathbf{y})$. Let Q denote the set of first κ vertices. The previous theorem implies:

$$\sum_e y_e \geq (1 - 3\epsilon)C(y, \tilde{K}) = \frac{1 - 3\epsilon}{\epsilon}\sum_{v \in Q} S_v(\mathbf{y})$$

where the final equality follows from the definition of $C(y, \tilde{K})$. Let $v^* \in V \setminus Q$ have the largest $S_v(\mathbf{y})$. Since $S_v(\mathbf{y}) \geq S_{v^*}(\mathbf{y})$ for all $v \in Q$ and since $|S_v(\mathbf{y})| = \kappa$, we have $\sum_{v \in Q} S_v(\mathbf{y}) \geq \kappa S_{v^*}(\mathbf{y})$.

Construct $\mathbf{z}$ as follows: For every e incident on a vertex in Q, we set $z_e = 0$; otherwise, set $z_e = y_e$. Note that $S_v(\mathbf{z}) \leq S_v(\mathbf{y})$ and $\max_v S_v(\mathbf{z}) = S_{v^*}(\mathbf{y})$. We have the following sequence of inequalities:

$$\sum_e z_e \geq \sum_e y_e - \sum_{v \in Q} S_v(\mathbf{y}) \geq \left(\frac{1 - 3\epsilon}{\epsilon} - 1\right)\sum_{v \in Q} S_v(\mathbf{y}) \geq \frac{1 - 4\epsilon}{\epsilon} \cdot \tilde{K} \cdot S_{v^*}(\mathbf{y}) = (1 - 4\epsilon)\tilde{K} S_{v^*}(\mathbf{y})$$

Since $\max_v S_v(\mathbf{z}) = S_{v^*}(\mathbf{y})$ by construction, we have:

$$\frac{\sum_e z_e}{\max_v S_v(\mathbf{z})} \geq (1 - 4\epsilon)\tilde{K} \geq (1 - 5\epsilon)K^*$$

for $\epsilon \in (0, 1/6)$. By scaling so that $\max_v S_v(\mathbf{z}) = 1$, it is easy to check that the new vector $\mathbf{z}$ is a feasible fractional matching with value at least $(1 - 5\epsilon)K^*$. We therefore have:

Theorem 10. *For $\epsilon \in (0, 1/6)$, a fractional matching of value $K^*(1 - \epsilon)$ can be computed in $O\left(m\frac{\log m}{\epsilon^3}\right)$ time.*

MAPREDUCE *Implementation.* The key difference in the ORACLE computation is that it requires summing the top $K\epsilon$ values $S_v(\mathbf{y})$. Since we assume reducer size $O(d_{max})$, the maximum degree which can be much smaller than n, this step will in general take $O(\log_{d_{max}} n)$ phases to execute (via random partitioning). Therefore, for any given K, the multiplicative weight procedure will require $O\left(\frac{\log^2 n}{\epsilon^3 \log d_{max}}\right)$ phases to execute. The remaining details are the same as before, so that the overall implementation takes $O\left(\frac{\log^2 n}{\epsilon^3 \log d_{max}}\right)$ MAPREDUCE phases with $O\left(m\frac{\log n}{\epsilon}\right)$ sequential complexity per phase and $O(d_{max})$ reducer size.

References

1. Afrati, F.N., Das Sarma, S., Salihoglu, S., Ullman, J.D.: Upper and lower bounds on the cost of a map-reduce computation. PVLDB **6**, 277–288 (2013)
2. Ahn, K.J., Guha, S.: Linear programming in the semi-streaming model with application to the maximum matching problem. In: Aceto, L., Henzinger, M., Sgall, J. (eds.) ICALP 2011, Part II. LNCS, vol. 6756, pp. 526–538. Springer, Heidelberg (2011)
3. Arora, S., Hazan, E., Kale, S.: The multiplicative weights update method: a meta algorithm and applications. Theory of Computing **8**, 121–164 (2012)
4. Awerbuch, B., Khandekar, R., Rao, S.: Distributed algorithms for multicommodity flow problems via approximate steepest descent framework. In: SODA, pp. 949–957 (2007)
5. Bahmani, B., Kumar, R., Vassilvitskii, S.: Densest subgraph in streaming and mapreduce. PVLDB **5**(5), 454–465 (2012)
6. Charikar, M.: Greedy approximation algorithms for finding dense components in a graph. In: Jansen, K., Khuller, S. (eds.) APPROX 2000. LNCS, vol. 1913, pp. 84–95. Springer, Heidelberg (2000)
7. Dean, J., Ghemawat, S.: Mapreduce: Simplified data processing on large clusters. In: OSDI, pp. 137–150 (2004)
8. Garg, N., Könemann, J.: Faster and simpler algorithms for multicommodity flow and other fractional packing problems. SIAM J. Comput. **37**(2), 630–652 (2007)
9. Grigoriadis, M.D., Khachiyan, L.G.: Approximate minimum-cost multicommodity flows in $\tilde{O}(\epsilon^{-2}knm)$ time. Math. Program. **75**, 477–482 (1996)
10. Apache hadoop, http://hadoop.apache.org
11. Kannan, R., Vinay, V.: Analyzing the structure of large graphs. Manuscript (1999)
12. Karloff, H., Suri, S., Vassilvitskii, S.: A model of computation for mapreduce. In: SODA (2010)
13. Motwani, R., Panigrahy, R., Xu, Y.: Fractional matching via balls-and-bins. In: Díaz, J., Jansen, K., Rolim, J.D.P., Zwick, U. (eds.) APPROX 2006 and RANDOM 2006. LNCS, vol. 4110, pp. 487–498. Springer, Heidelberg (2006)
14. Plotkin, S.A., Shmoys, D.B., Tardos, É.:. Fast approximation algorithms for fractional packing and covering problems. In: FOCS, pp. 495–504 (1991)
15. Saha, B., Hoch, A., Khuller, S., Raschid, L., Zhang, X.-N.: Dense subgraphs with restrictions and applications to gene annotation graphs. In: Berger, B. (ed.) RECOMB 2010. LNCS, vol. 6044, pp. 456–472. Springer, Heidelberg (2010)
16. Suri, S., Vassilvitskii, S.: Counting triangles and the curse of the last reducer. In: WWW, pp. 607–614 (2011)
17. Young, N.E.: Randomized rounding without solving the linear program. In: SODA (1995)

Computing Diffusion State Distance Using Green's Function and Heat Kernel on Graphs

Edward Boehnlein[1], Peter Chin[2], Amit Sinha[2], and Linyuan Lu[3]([✉])

[1] University of South Carolina, Columbia, SC 29208, USA
`boehnlei@email.sc.edu`
[2] Boston University, Boston, MA 02215, USA
`spchin@cs.bu.edu, amits@bu.edu`
[3] University of South Carolina, Columbia, SC 29208, USA
`lu@math.sc.edu`

Abstract. The diffusion state distance (DSD) was introduced by Cao-Zhang-Park-Daniels-Crovella-Cowen-Hescott [*PLoS ONE, 2013*] to capture functional similarity in protein-protein interaction networks. They proved the convergence of DSD for non-bipartite graphs. In this paper, we extend the DSD to bipartite graphs using lazy-random walks and consider the general L_q-version of DSD. We discovered the connection between the DSD L_q-distance and Green's function, which was studied by Chung and Yau [*J. Combinatorial Theory (A), 2000*]. Based on that, we computed the DSD L_q-distance for Paths, Cycles, Hypercubes, as well as random graphs $G(n,p)$ and $G(w_1, \ldots, w_n)$. We also examined the DSD distances of two biological networks.

1 Introduction

Recently, the diffusion state distance (DSD, for short) was introduced in [3] to capture functional similarity in protein-protein interaction (PPI) networks. The diffusion state distance is much more effective than the classical shortest-path distance for the problem of transferring functional labels across nodes in PPI networks, based on evidence presented in [3]. The definition of DSD is purely graph theoretic and is based on random walks.

Let $G = (V, E)$ be a simple undirected graph on the vertex set $\{v_1, v_2, \ldots, v_n\}$. For any two vertices u and v, let $He^{\{k\}}(u, v)$ be the expected number of times that a random walk starting at node u and proceeding for k steps, will visit node v. Let $He^{\{k\}}(u)$ be the vector $(He^{\{k\}}(u, v_1), \ldots, He^{\{k\}}(u, v_n))$. The diffusion state distance (or DSD, for short) between two vertices u and v is defined as

$$DSD(u, v) = \lim_{k \to \infty} \left\| He^{\{k\}}(u) - He^{\{k\}}(v) \right\|_1$$

Peter Chin: Supported in part by NSF grant DMS 1222567 as well as AFOSR grant FA9550-12-1-0136.
Linyuan Lu: Research supported in part by NSF grant DMS 1300547 and ONR grant N00014-13-1-0717.

© Springer International Publishing Switzerland 2014
A. Bonato et al. (Eds.): WAW 2014, LNCS 8882, pp. 79–95, 2014.
DOI: 10.1007/978-3-319-13123-8_7

provided the limit exists (see [3]). Here the L_1-norm is not essential. Generally, for $q \geq 1$, one can define the DSD L_q-distance as

$$DSD_q(u, v) = \lim_{k \to \infty} \left\| He^{\{k\}}(u) - He^{\{k\}}(v) \right\|_q$$

provided the limit exists. (We use L_q rather than L_p to avoid confusion, as p will be used as a probability throughout the paper.)

In [3], Cowen et al. showed that the above limit always exists whenever the random walk on G is ergodic (i.e., G is connected non-bipartite graph). They also prove that this distance can be computed by the following formula:

$$DSD(u, v) = \left\| (1_u - 1_v)(I - D^{-1}A + W)^{-1} \right\|_1$$

where D is the diagonal degree matrix, A is the adjacency matrix, and W is the constant matrix in which each row is a copy of π, $\pi = \frac{1}{\sum_{i=1}^{n} d_i}(d_1, \ldots, d_n)$ is the unique steady state distribution.

A natural question is how to define the diffusion state distance for a bipartite graph. We suggest to use the lazy random walk. For a given $\alpha \in (0, 1)$, one can choose to stay at the current node u with probability α, and choose to move to one of its neighbors with probability $(1 - \alpha)/d_u$. In other words, the transitive matrix of the α-lazy random walk is

$$T_\alpha = \alpha I + (1 - \alpha)D^{-1}A.$$

Similarly, let $He_\alpha^{\{k\}}(u, v)$ be the expected number of times that the α-lazy random walk starting at node u and proceeding for k steps, will visit node v. Let $He_\alpha^{\{k\}}(u)$ be the vector $(He_\alpha^{\{k\}}(u, v_1), \ldots, He_\alpha^{\{k\}}(u, v_n))$. The α-diffusion state distance L_q-distance between two vertices u and v is

$$DSD_q^\alpha(u, v) = \lim_{k \to \infty} \left\| He_\alpha^{\{k\}}(u) - He_\alpha^{\{k\}}(v) \right\|_q.$$

Theorem 1. *For any connected graph G and $\alpha \in (0, 1)$, the $DSD_q^\alpha(u, v)$ is always well-defined and satisfies*

$$DSD_q^\alpha(u, v) = (1 - \alpha)^{-1} \left\| (1_u - 1_v)\mathbb{G} \right\|_q. \tag{1}$$

Here $\mathbb{G}$ is the matrix of Green's function of G.

Observe that $(1 - \alpha)DSD_q^\alpha(u, v)$ is independent of the choice of α. Naturally, we define the DSD L_q-distance of any graph G as:

$$DSD_q(u, v) := \lim_{\alpha \to 0} (1 - \alpha)DSD_q^\alpha(u, v) = \left\| (1_u - 1_v)\mathbb{G} \right\|_q.$$

This definition extends the original definition for non-bipartite graphs.

With properly chosen α, $\left\| He_\alpha^{\{k\}}(u) - He_\alpha^{\{k\}}(v) \right\|_q$ converges faster than $\left\| He^{\{k\}}(u) - He^{\{k\}}(v) \right\|_q$. This fact leads to a faster algorithm to estimate a single distance $DSD_q(u, v)$ using random walks. We will discuss it in Remark 1.

Green's function was introduced in 1828 by George Green [17] to solve some partial differential equations, and it has found many applications (e.g. [1], [5],[9], [16], [19], [24]).

The Green's function on graphs was first investigated by Chung and Yau [5] in 2000. Given a graph $G = (V, E)$ and a given function $g \colon V \to \mathbb{R}$, consider the problem to find f satisfying the discrete Laplace equation

$$Lf = \sum_{y \in V} (f(x) - f(y)) p_{xy} = g(x).$$

Here p_{xy} is the transition probability of the random walk from x to y. Roughly speaking, Green's function is the left inverse operator of L (for the graphs with boundary). It is closely related to the Heat kernel of the graphs (see also [15]) and the normalized Laplacian.

In this paper, we will use Green's function to compute the DSD L_q-distance for various graphs. The maximum DSD L_q-distance varies from graphs to graphs. The maximum DSD L_q-distance for paths and cycles are at the order of $\Theta(n^{1+1/q})$ while the L_q-distance for some random graphs $G(n, p)$ and $G(w_1, \ldots, w_n)$ are constant for some ranges of p. The hypercubes are somehow between the two classes. The DSD L_1-distance is $\Omega(n)$ while the L_q-distance is $\Theta(1)$ for $q > 1$. Our method for random graphs is based on the strong concentration of the Laplacian eigenvalues.

The paper is organized as follows. In Section 2, we will briefly review the terminology on the Laplacian eigenvalues, Green's Function, and heat kernel. The proof of Theorem 1 will be proved in Section 3. In Section 4, we apply Green's function to calculate the DSD distance for various symmetric graphs like paths, cycles, and hypercubes. We will calculate the DSD L_2-distance for random graphs $G(n, p)$ and $G(w_1, w_2, \ldots, w_n)$ in Section 5. In the last section, we examined two brain networks: a cat and a Rhesus monkey. The distributions of the DSD distances are calculated.

2   Notation and Background

In this paper, we only consider undirected simple graph $G = (V, E)$ with the vertex set V and the edge set E. For each vertex $x \in V$, the *neighborhood* of x, denoted by $N(x)$, is the set of vertices adjacent to x. The *degree* of x, denoted by d_x, is the cardinality of $N(x)$. We also denote the maximum degree by Δ and the minimum degree by δ.

Without loss of generalization, we assume that the set of vertices is ordered and assume $V = [n] = \{1, 2, \ldots, n\}$. Let A be the adjacency matrix and $D = \mathrm{diag}(d_1, \ldots, d_n)$ be the diagonal matrix of degrees. For a given subset S, let the volume of S to be $\mathrm{vol}(S) := \sum_{i \in S} d_i$. In particular, we write $\mathrm{vol}(G) = \mathrm{vol}(V) = \sum_{i=1}^{n} d_i$.

Let V^* be the linear space of all real functions on V. The *discrete Laplace operator* $L\colon V^* \to V^*$ is defined as

$$L(f)(x) = \sum_{y \in N(x)} \frac{1}{d_x}(f(x) - f(y)).$$

The Laplace operator can also written as a $(n \times n)$-matrix:

$$L = I - D^{-1}A.$$

Here $D^{-1}A$ is the transition probability matrix of the (uniform) random walk on G. Note that L is not symmetric. We consider a symmetric version

$$\mathcal{L} := I - D^{-1/2}AD^{-1/2} = D^{1/2}LD^{-1/2},$$

which is so called the *normalized Laplacian*. Both L and $\mathcal{L}$ have the same set of eigenvalues. The eigenvalues of $\mathcal{L}$ can be listed as

$$0 = \lambda_0 \leq \lambda_1 \leq \lambda_2 \leq \cdots \leq \lambda_{n-1} \leq 2.$$

The eigenvalue $\lambda_1 > 0$ if and only if G is connected while $\lambda_{n-1} = 2$ if and only if G is a bipartite graph. Let $\phi_0, \phi_1, \ldots, \phi_{n-1}$ be a set of orthogonal unit eigenvectors. Here $\phi_0 = \frac{1}{\sqrt{\mathrm{vol}(G)}}(\sqrt{d_1}, \ldots, \sqrt{d_n})$ is the positive unit eigenvector for $\lambda_0 = 0$ and ϕ_i is the eigenvector for λ_i $(1 \leq i \leq n-1)$.

Let $O = (\phi_0, \ldots, \phi_{n-1})$ and $\Lambda = \mathrm{diag}(0, \lambda_1, \ldots, \lambda_{n-1})$. Then O is an orthogonal matrix and $\mathcal{L}$ be diagonalized as

$$\mathcal{L} = O\Lambda O'. \tag{2}$$

Equivalently, we have

$$L = D^{-1/2}O\Lambda O'D^{1/2}. \tag{3}$$

The *Green's function* $\mathbb{G}$ is the matrix with its entries, indexed by vertices x and y, defined by a set of two equations:

$$\mathbb{G}L(x, y) = I(x, y) - \frac{d_y}{\mathrm{vol}(G)}, \tag{4}$$

$$\mathbb{G}\mathbf{1} = 0. \tag{5}$$

(This is the so-called Green's function for graphs without boundary in [5].)

The *normalized Green's function* $\mathcal{G}$ is defined similarly:

$$\mathcal{G}\mathcal{L}(x, y) = I(x, y) - \frac{\sqrt{d_x d_y}}{\mathrm{vol}(G)}.$$

The matrices $\mathbb{G}$ and $\mathcal{G}$ are related by

$$\mathcal{G} = D^{1/2}\mathbb{G}D^{-1/2}.$$

Alternatively, $\mathcal{G}$ can be defined using the eigenvalues and eigenvectors of $\mathcal{L}$ as follows:

$$\mathcal{G} = O\Lambda^{\{-1\}}O',$$

where $\Lambda^{\{-1\}} = \mathrm{diag}(0, \lambda_1^{-1}, \ldots, \lambda_{n-1}^{-1})$. Thus, we have

$$G(x,y) = \sum_{l=1}^{n-1} \frac{1}{\lambda_l} \sqrt{\frac{d_y}{d_x}} \phi_l(x)\phi_l(y). \tag{6}$$

For any real $t \geq 0$, the heat kernel $\mathcal{H}_t$ is defined as

$$\mathcal{H}_t = e^{-t\mathcal{L}}.$$

Thus,

$$\mathcal{H}_t(x,y) = \sum_{l=0}^{n-1} e^{-\lambda_i t} \phi_l(x)\phi_l(y).$$

The heat kernel $\mathcal{H}_t$ satisfies the heat equation

$$\frac{d}{dt}\mathcal{H}_t f = -\mathcal{L}\mathcal{H}_t f.$$

The relation of the heat kernel and Green's function is given by

$$\mathcal{G} = \int_0^\infty \mathcal{H}_t dt - \phi_0'\phi_0.$$

The heat kernel can be used to compute Green's function for the Cartesian product of two graphs. We will omit the details here. Readers are directed to [5] and [6] for the further information.

3   Proof of Main Theorem

Proof (Proof of Theorem 1:). Rewrite the transition probability matrix T_α as

$$\begin{aligned} T_\alpha &= \alpha I + (1-\alpha)D^{-1}A. \\ &= D^{-1/2}(\alpha I + (1-\alpha)D^{-1/2}AD^{-1/2})D^{1/2} \\ &= D^{-1/2}(\alpha I + (1-\alpha)(I - \mathcal{L}))D^{1/2} \\ &= D^{-1/2}(I - (1-\alpha)\mathcal{L})D^{1/2}. \end{aligned}$$

For $k = 0, 1, \ldots, n-1$, let $\lambda_k^* = 1 - (1-\alpha)\lambda_k$ and $\Lambda^* = diag(\lambda_0^*, \ldots, \lambda_{n-1}^*) = I - (1-\alpha)\Lambda$. Applying Equation (3), we get

$$T_\alpha = D^{-1/2}O\Lambda^*O'D^{1/2} = (O'D^{1/2})^{-1}\Lambda^*O'D^{1/2}.$$

Then for any $t \geq 1$, the t-step transition matrix is $T_\alpha^t = (OD^{1/2})^{-1}\Lambda^{*t}OD^{1/2} = D^{-1/2}O\Lambda^{*t}O'D^{1/2}$. Denote $p_\alpha^{\{t\}}(u, j)$ as the $(u, j)^{th}$ entry in T_α^t.

$$p_\alpha^{\{t\}}(u, j) = \sum_{l=0}^{n-1}(\lambda_l^*)^t\sqrt{\frac{d_j}{d_u}}\phi_l(u)\phi_l(j)$$

$$= \frac{d_j}{vol(G)} + \sum_{l=1}^{n-1}(\lambda_l^*)^t\sqrt{\frac{d_j}{d_u}}\phi_l(u)\phi_l(j).$$

Thus,

$$He_\alpha^{\{k\}}(u, j) - He_\alpha^{\{k\}}(v, j) = \sum_{t=0}^{k}\sum_{l=1}^{n-1}(\lambda_l^*)^t d_j^{1/2}\phi_l(j)(d_u^{-1/2}\phi_l(u) - d_v^{-1/2}\phi_l(v)).$$

The limit $\lim_{k\to\infty} He_\alpha^{\{k\}}(u, j) - He_\alpha^{\{k\}}(v, j)$ forms the sum of n geometric series:

$$\sum_{t=0}^{\infty}\sum_{l=1}^{n-1}(\lambda_l^*)^t d_j^{1/2}\phi_l(j)(d_u^{-1/2}\phi_l(u) - d_v^{-1/2}\phi_l(v)).$$

Note each geometric series converges since the common ratio $\lambda_l^* \in (-1, 1)$. Thus,

$$\lim_{k\to\infty}\left(He_\alpha^{\{k\}}(u, j) - He_\alpha^{\{k\}}(v, j)\right) = \sum_{t=0}^{\infty}\sum_{l=1}^{n-1}(\lambda_l^*)^t d_j^{1/2}\phi_l(j)(d_u^{-1/2}\phi_l(u) - d_v^{-1/2}\phi_l(v))$$

$$= \sum_{l=1}^{n-1} d_j^{1/2}\phi_l(j)(d_u^{-1/2}\phi_l(u) - d_v^{-1/2}\phi_l(v))\sum_{t=0}^{\infty}(\lambda_l^*)^t$$

$$= \sum_{l=1}^{n-1}\frac{1}{1-\lambda_l^*}d_j^{1/2}\phi_l(j)(d_u^{-1/2}\phi_l(u) - d_v^{-1/2}\phi_l(v))$$

$$= \frac{1}{1-\alpha}\sum_{l=1}^{n-1}\frac{1}{\lambda_l}d_j^{1/2}\phi_l(j)(d_u^{-1/2}\phi_l(u) - d_v^{-1/2}\phi_l(v))$$

$$= \frac{1}{1-\alpha}(\mathbb{G}(u, j) - \mathbb{G}(v, j)).$$

We have

$$\lim_{k\to\infty} He_\alpha^{\{k\}}(u) - He_\alpha^{\{k\}}(v) = \frac{1}{1-\alpha}(\mathbf{1}_u - \mathbf{1}_v)\mathbb{G}.$$

Remark 1. Observe that the convergence rate of $He_\alpha^{\{k\}}(u) - He_\alpha^{\{k\}}(v)$ is determined by $\bar{\lambda}^* := \max\{1 - (1-\alpha)\lambda_1, (1-\alpha)\lambda_{n-1} - 1)$. It is critical that we assume $\alpha \neq 0$. When $\alpha = 0$ then $\bar{\lambda}^* < 1$ holds only if $\lambda_{n-1} < 2$, i.e. G is a non-bipartite graph (see [3]).

When $\lambda_1 + \lambda_{n-1} > 2$, $\bar{\lambda}^*$ (as a function of α) achieves the minimum value $\frac{\lambda_{n-1}-\lambda_1}{\lambda_{n-1}+\lambda_1}$ at $\alpha = 1 - \frac{2}{\lambda_1+\lambda_{n-1}}$. This is the best mixing rate that the α-lazy random walk on G can achieve. Using the α-lazy random walks (with $\alpha = 1 - \frac{2}{\lambda_1+\lambda_{n-1}}$) to approximate the DSD L_q-distance will be faster than using regular random walks.

Equation (6) implies $\|\mathbb{G}\|_2 \leq \frac{1}{\lambda_1}\sqrt{\frac{\Delta}{\delta}}$. Combining with Theorem 1, we have

Corollary 1. *For any connected simple graph G, and any two vertices u and v, we have* $DSD_2(u,v) \leq \frac{\sqrt{2}}{\lambda_1}\sqrt{\frac{\Delta}{\delta}}$.

Note that for any connected graph G with diameter m (Lemma 1.9, [6])

$$\lambda_1 > \frac{1}{m\,\mathrm{vol}(G)}.$$

This implies a uniform bound for the DSD L_2 distances on any connected graph G on n vertices.

$$DSD_2(u,v) \leq \sqrt{\frac{2\Delta}{\delta}}\,m\,\mathrm{vol}(G) < \sqrt{2}n^{3.5}.$$

This is a very coarse upper bound. But it does raise an interesting question "How large can the DSD L_q-distance be?"

4   Some Examples of the DSD Distance

In this section, we use Green's function to compute the DSD L_q-distance (between two vertices of the distance reaching the diameter) for paths, cycles, and hypercubes.

4.1   The Path P_n

We label the vertices of P_n as $1, 2, \ldots, n$, in sequential order. Chung and Yau computed the Green's function $\mathcal{G}$ of the weighed path with no boundary (Theorem 9, [5]). It implies that Green's function of the path P_n is given by: for any $u \leq v$,

$$
\begin{aligned}
\mathcal{G}(u,v) &= \frac{\sqrt{d_u d_v}}{4(n-1)^2}\left(\sum_{z<u}(d_1+\ldots+d_z)^2 + \sum_{v\leq z}(d_{z+1}+\cdots+d_n)^2\right.\\
&\qquad\left. - \sum_{u\leq z<v}(d_1+\cdots+d_z)(d_{z+1}+\cdots+d_n)\right)\\
&= \frac{\sqrt{d_u d_v}}{4(n-1)^2}\left(\sum_{z=1}^{u-1}(2z-1)^2 + \sum_{z=v}^{n-1}(2n-2z-1)^2 - \sum_{z=u}^{v-1}(2z-1)(2n-2z-1)\right)\\
&= \frac{\sqrt{d_u d_v}}{4(n-1)^2}\left(\sum_{z=1}^{n-1}(2z-1)^2 + \sum_{z=v}^{n-1}(2n-2)(2n-4z) - \sum_{z=u}^{v-1}(2z-1)(2n-2)\right)\\
&= \frac{\sqrt{d_u d_v}(2n-1)(2n-3)}{12(n-1)} + \frac{\sqrt{d_u d_v}}{2(n-1)}\left(\sum_{z=v}^{n-1}(2n-4z) - \sum_{z=u}^{v-1}(2z-1)\right)\\
&= \frac{\sqrt{d_u d_v}}{2(n-1)}\left((u-1)^2 + (n-v)^2 - \frac{2n^2-4n+3}{6}\right).
\end{aligned}
$$

When $u > v$, we have

$$\mathcal{G}(u,v) = \mathcal{G}(v,u) = \frac{\sqrt{d_u d_v}}{2(n-1)}\left((v-1)^2 + (n-u)^2 - \frac{2n^2 - 4n + 3}{6}\right).$$

Applying $\mathbb{G}(u,v) = \frac{\sqrt{d_v}}{\sqrt{d_u}}\mathcal{G}(u,v)$, we get

$$\mathbb{G}(u,v) = \begin{cases} \frac{d_v}{2(n-1)}\left((u-1)^2 + (n-v)^2 - \frac{2n^2-4n+3}{6}\right) & \text{if } u \le v; \\ \frac{d_v}{2(n-1)}\left((v-1)^2 + (n-u)^2 - \frac{2n^2-4n+3}{6}\right) & \text{if } u > v. \end{cases}$$

We have

$$\mathbb{G}(1,1) = \frac{4n^2 - 8n + 3}{12(n-1)};$$

$$\mathbb{G}(1,j) = \frac{1}{n-1}\left((n-j)^2 - \frac{2n^2 - 4n + 3}{6}\right) \qquad \text{for } 2 \le j \le n-1;$$

$$\mathbb{G}(1,n) = -\frac{2n^2 - 4n + 3}{12(n-1)};$$

$$\mathbb{G}(n,1) = -\frac{2n^2 - 4n + 3}{12(n-1)};$$

$$\mathbb{G}(n,j) = \frac{1}{n-1}\left((j-1)^2 - \frac{2n^2 - 4n + 3}{6}\right) \qquad \text{for } 2 \le j \le n-1;$$

$$\mathbb{G}(n,n) = \frac{4n^2 - 8n + 3}{12(n-1)}.$$

Thus,

$$\mathbb{G}(1,j) - \mathbb{G}(n,j) = \begin{cases} \frac{n-1}{2} & \text{if } j = 1; \\ n+1-2j & \text{if } 2 \le j \le n-1; \\ -\frac{n-1}{2} & \text{if } j = n. \end{cases} \qquad (7)$$

Theorem 2. *For any $q \ge 1$, the DSD L_q-distance of the Path P_n between 1 and n satisfies*

$$DSD_q(1,n) = (1+q)^{-1/q}n^{1+1/q} + O(n^{1/q}).$$

Proof.

$$DSD_q(1,n) = \left(2\left(\frac{n-1}{2}\right)^q + \sum_{j=2}^{n-1}|n+1-2j|^q\right)^{1/q}$$

$$= \left(\frac{1}{1+q}n^{1+q} + O(n^q)\right)^{1/q}$$

$$= (1+q)^{-1/q}n^{1+1/q} + O(n^{1/q}).$$

For $q = 1$, we have the following exact result:

$$DSD_1(1, n) = \sum_{j=1}^{n} |\mathbb{G}(1, j) - \mathbb{G}(n, j)|$$

$$= \begin{cases} 2k^2 - 2k + 1 & \text{if } n = 2k \\ 2k^2 & \text{if } n = 2k + 1. \end{cases}$$

4.2   The Cycle C_n

Now we consider Green's function of cycle C_n. For $x, y \in \{1, 2, \ldots, n\}$, let $|x - y|_c$ be the graph distance of x, y in C_n. We have the following Lemma.

Lemma 1. *For even $n = 2k$, Green's function $\mathbb{G}$ of C_n is given by*

$$\mathbb{G}(x, y) = \frac{1}{2k}(k - |x - y|_c)^2 - \frac{k}{6} - \frac{1}{12k}.$$

For odd $n = 2k + 1$, Green's function $\mathbb{G}$ of C_n is given by

$$\mathbb{G}(x, y) = \frac{2}{2k + 1}\left(\frac{k + 1 - |x - y|_c}{2}\right) - \frac{k^2 + k}{3(2k + 1)}.$$

Proof. We only prove the even case here. The odd case is similar and will be left to the readers.

For $n = 2k$, it suffices to verify that $\mathbb{G}$ satisfies Equations (4) and (5). To verify Equation (4), we need show

$$\mathbb{G}(x, y) - \frac{1}{2}\mathbb{G}(x, y - 1) - \frac{1}{2}\mathbb{G}(x, y + 1) = \begin{cases} -\frac{1}{n} & \text{if } x \neq y; \\ 1 - \frac{1}{n} & \text{if } x = y. \end{cases}$$

Let $z = \frac{k}{6} + \frac{1}{12k}$ and $i = |x - y|_c$. For $x \neq y$, we have

$$\mathbb{G}(x, y) - \frac{1}{2}\mathbb{G}(x, y - 1) - \frac{1}{2}\mathbb{G}(x, y + 1)$$

$$= (\frac{1}{2k}(k - i)^2 - z) - \frac{1}{2}(\frac{1}{2k}(k - i - 1)^2 - z) - \frac{1}{2}(\frac{1}{2k}(k - i + 1)^2 - z)$$

$$= -\frac{1}{2k}$$

$$= -\frac{1}{n}.$$

When $x = y$, we have

$$\mathbb{G}(x, y) - \frac{1}{2}\mathbb{G}(x, y - 1) - \frac{1}{2}\mathbb{G}(x, y + 1)$$

$$= \frac{1}{2k}k^2 - z - \frac{1}{2}\left(\frac{1}{2k}(k-1)^2 - z\right) - \frac{1}{2}\left(\frac{1}{2k}(k-1)^2 - z\right)$$

$$= \frac{2k-1}{2k}$$

$$= 1 - \frac{1}{n}.$$

To verify Equation (5), it is enough to verify

$$1^2 + 2^2 + \cdots + (k-1)^2 + k^2 + (k-1)^2 + \cdots + 1^2 = \frac{2k^3 + k}{3} = n^2 z.$$

This can be done by induction on k.

Theorem 3. *For any $q \geq 1$, the DSD L_q-distance of the Cycle C_n between 1 and $\lfloor \frac{n}{2} \rfloor + 1$ satisfies*

$$DSD_q(1, \lfloor\frac{n}{2}\rfloor + 1) = \left(\frac{4}{1+q}\right)^{1/q}\left(\frac{n}{4}\right)^{1+1/q} + O(n^{1/q}).$$

Proof. We only verify the case of even cycle here. The odd cycle is similar and will be omitted.

For $n = 2k$, the difference of $\mathbb{G}(1, j)$ and $\mathbb{G}(1 + k, j)$ have a simple form:

$$\mathbb{G}(1, j) - \mathbb{G}(1 + k, j) = \frac{1}{2k}((k-i)^2 - i^2) = \frac{k}{2} - i,$$

where $i = |j - 1|_c$. Thus,

$$DSD_q(1, 1 + k) = \left(2\sum_{i=0}^{k-1}\left|\frac{k}{2} - i\right|^q\right)^{1/q}$$

$$= \left(\frac{4}{1+q}\left(\frac{k}{2}\right)^{1+q} + O(k^q)\right)^{1/q}$$

$$= \left(\frac{4}{1+q}\right)^{1/q}\left(\frac{n}{4}\right)^{1+1/q} + O(n^{1/q}).$$

4.3   The Hypercube Q_n

Now we consider the hypercube Q_n, whose vertices are the binary strings of length n and whose edges are pairs of vertices differing only at one coordinate.

Chung and Yau [5] computed the Green's function of Q_n: for any two vertices x and y with distance k in Q_n,

$$\mathbb{G}(x,y) = 2^{-2n}\left(-\sum_{j<k}\frac{(\binom{n}{0}+\cdots+\binom{n}{j})(\binom{n}{j+1}+\cdots+\binom{n}{n})}{\binom{n-1}{j}} + \sum_{k\le j}\frac{(\binom{n}{j+1}+\cdots+\binom{n}{n})^2}{\binom{n-1}{j}}\right)$$

$$= 2^{-2n}\sum_{j=0}^{n}\frac{(\binom{n}{j+1}+\cdots+\binom{n}{n})^2}{\binom{n-1}{j}} - 2^{-n}\sum_{j<k}\frac{\binom{n}{j+1}+\cdots+\binom{n}{n}}{\binom{n-1}{j}}.$$

We are interested in the DSD distance between a pair of antipodal vertices. Let $\mathbf{0}$ denote the all-0-string and $\mathbf{1}$ denote the all-1-string. For any vertex x, if the distance between $\mathbf{0}$ and x is i then the distance between $\mathbf{1}$ and x is $n - i$. We have

$$\mathbb{G}(\mathbf{0},x) - \mathbb{G}(\mathbf{1},x) = -2^{-n}\sum_{j<k}\frac{\binom{n}{j+1}+\cdots+\binom{n}{n}}{\binom{n-1}{j}} + 2^{-n}\sum_{j<n-k}\frac{\binom{n}{j+1}+\cdots+\binom{n}{n}}{\binom{n-1}{j}}$$

$$= 2^{-n}\sum_{j=k}^{n-k-1}\frac{\binom{n}{j+1}+\cdots+\binom{n}{n}}{\binom{n-1}{j}}. \tag{8}$$

Here we use the convention that $\sum_{j=b}^{a} c_j = -\sum_{j=a}^{b} c_j$ for $b > a$.

Theorem 4. *For any $q \ge 1$, the DSD L_q-distance of the hypercube Q_n between $\mathbf{0}$ and $\mathbf{1}$ satisfies*

$$DSD_q(\mathbf{0},\mathbf{1}) = \left(\sum_{k=0}^{n}\binom{n}{k}\left|2^{-n}\sum_{j=k}^{n-k-1}\frac{\binom{n}{j+1}+\cdots+\binom{n}{n}}{\binom{n-1}{j}}\right|^q\right)^{1/q}. \tag{9}$$

In particular, $DSD_q(\mathbf{0},\mathbf{1}) = \Theta(1)$ when $q > 1$ while $DSD_1(\mathbf{0},\mathbf{1}) = \Omega(n)$.

Proof. Equation (9) follows from the definition of DSD L_q-distance and Equation (8). Let

$$a_k = \binom{n}{k}\left|2^{-n}\sum_{j=k}^{n-k-1}\frac{\binom{n}{j+1}+\cdots+\binom{n}{n}}{\binom{n-1}{j}}\right|^q.$$

Observe that $a_k = a_{n-k}$, we only need to estimate a_k for $0 \le k \le n/2$. Also we can throw away the terms in the second summation for $j > n/2$ since that part is at most half of a_k. For $k \le j \le n/2$,

$$\frac{1}{2} \le 2^{-n}\left(\binom{n}{j+1}+\cdots+\binom{n}{n}\right) \le 1.$$

Thus a_k has the same magnitude as $b_k := \binom{n}{k}\left(\sum_{j=k}^{n/2}\frac{1}{\binom{n-1}{j}}\right)^q.$

For $q > 1$, we first bound b_k by $b_k \leq \binom{n}{k} \left(\frac{n/2}{\binom{n-1}{k}} \right)^q = O(n^{(1-q)k+q})$. When $k > \frac{q+2}{q-1}$, we have $b_k = O(n^{-2})$. The total contribution of those b_k's is $O(n^{-1})$, which is negligible. Now consider the term b_k for $k = 0, 1, \ldots, \lfloor \frac{q+2}{q-1} \rfloor$. We bound b_k by

$$b_k \leq \binom{n}{k} \left(\frac{1}{\binom{n-1}{k}} + \frac{n/2}{\binom{n-1}{k+1}} \right)^q = O(1).$$

This implies $DSD_q(\mathbf{0}, \mathbf{1}) = O(1)$. The lower bound $DSD_q(\mathbf{0}, \mathbf{1}) \geq 1$ is obtained by taking the term at $k = 0$. Putting together, we have $DSD_q(\mathbf{0}, \mathbf{1}) = \Theta(1)$ for $q > 1$.

For $q = 1$, note that

$$b_k = \sum_{j=k}^{n/2} \frac{\binom{n}{k}}{\binom{n-1}{j}} > \frac{\binom{n}{k}}{\binom{n-1}{k}} = \frac{n}{n-k} > 1.$$

Thus, $DSD_1(\mathbf{0}, \mathbf{1}) = \Omega(n)$.

5   Random Graphs

In this section, we will calculate the DSD L_q-distance in two random graphs models. For random graphs, the non-zero Laplacian eigenvalues of a graph G are often concentrated around 1. The following Lemma is useful to the DSD L_q-distance.

Lemma 2. *Let $\lambda_1, \ldots, \lambda_{n-1}$ be all non-zero Laplacian eigenvalues of a graph G. Suppose there is a small number $\epsilon \in (0, 1/2)$, so that for $1 \leq i \leq n-1$, $|1 - \lambda_i| \leq \epsilon$. Then for any pairs of vertices u, v, the DSD L_q-distance satisfies*

$$|DSD_q(u, v) - 2^{1/q}| \leq \frac{\epsilon}{1-\epsilon} \sqrt{\frac{\Delta}{d_u} + \frac{\Delta}{d_v}} \quad if \ q \geq 2, \quad (10)$$

$$|DSD_q(u, v) - 2^{1/q}| \leq n^{\frac{1}{q} - \frac{1}{2}} \frac{\epsilon}{1-\epsilon} \sqrt{\frac{\Delta}{d_u} + \frac{\Delta}{d_v}} \quad for \ 1 \leq q < 2. \quad (11)$$

Proof. Rewrite the normalized Green's function $\mathcal{G}$ as

$$\mathcal{G} = I - \phi_0' \phi_0 + \Upsilon.$$

Note that the eigenvalues of $\Upsilon := \mathcal{G} - I + \phi_0 \phi_0'$ are $0, \frac{1}{\lambda_1} - 1, \ldots, \frac{1}{\lambda_{n-1}} - 1$. Observe that for each $i = 1, 2, \ldots, n-1$, $|\frac{1}{\lambda_i} - 1| \leq \frac{\epsilon}{1-\epsilon}$. We have

$$\|\Upsilon\| \leq \frac{\epsilon}{1-\epsilon}.$$

Thus,

$$
\begin{aligned}
\mathrm{DSD}_q(u, v) &= \|(\mathbf{1}_u - \mathbf{1}_v)D^{-1/2}\mathcal{G}D^{1/2}\|_q \\
&= \|(\mathbf{1}_u - \mathbf{1}_v)D^{-1/2}(I - \phi_0'\phi + \Upsilon)D^{1/2}\|_q \\
&\leq \|(\mathbf{1}_u - \mathbf{1}_v)D^{-1/2}(I - \phi_0'\phi)D^{1/2}\|_q + \|(\mathbf{1}_u - \mathbf{1}_v)D^{-1/2}\Upsilon D^{1/2}\|_q.
\end{aligned}
$$

Viewing Υ as the error term, we first calculate the main term.

$$
\begin{aligned}
\|(\mathbf{1}_u - \mathbf{1}_v)D^{-1/2}(I - \phi_0'\phi)D^{1/2}\|_q \\
= \|(\mathbf{1}_u - \mathbf{1}_v)(I - W)\|_q \\
= \|(\mathbf{1}_u - \mathbf{1}_v)\|_q \\
= 2^{1/q}.
\end{aligned}
$$

The L_2-norm of the error term can be bounded by

$$
\begin{aligned}
\|(\mathbf{1}_u - \mathbf{1}_v)D^{-1/2}\Upsilon D^{1/2}\|_2 \\
\leq \|(\mathbf{1}_u - \mathbf{1}_v)D^{-1/2}\|_2 \|\Upsilon\| \|D^{1/2}\| \\
\leq \sqrt{\frac{1}{d_u} + \frac{1}{d_v}} \frac{\epsilon}{1 - \epsilon} \sqrt{\Delta} \\
= \frac{\epsilon}{1 - \epsilon} \sqrt{\frac{\Delta}{d_u} + \frac{\Delta}{d_v}}.
\end{aligned}
$$

To get the bound of L_q-norm from L_2-norm, we apply the following relation of L_q-norm and L_2-norm to the error term. For any vector $x \in \mathbb{R}^n$,

$$
\|x\|_q \leq \|x\|_2 \quad \text{for } q \geq 2.
$$

and

$$
\|x\|_q \leq n^{\frac{1}{q} - \frac{1}{2}} \|x\|_2 \quad \text{for } 1 \leq q < 2.
$$

The inequalities (10) and (11) follow from the triangular inequality of the L_q-norm and the upper bound of the error term.

Now we consider the classical Erdős-Renyi random graphs $G(n, p)$. For a given n and $p \in (0, 1)$, $G(n, p)$ is a random graph on the vertex set $\{1, 2, \ldots, n\}$ obtained by adding each pair (i, j) to the edges of $G(n, p)$ with probability p independently.

There are plenty of references on the concentration of the eigenvalues of $G(n, p)$ (for example, [12], [14],[21], and [22]). Here we list some facts on $G(n, p)$.

1. For $p > \frac{(1+\epsilon)\log n}{n}$, almost surely $G(n, p)$ is connected.
2. For $p \gg \frac{\log n}{n}$, $G(n, p)$ is "almost regular"; namely for all vertex v, $d_v = (1 + o_n(1))np$.
3. For $np(1 - p) \gg \log^4 n$, all non-zero Laplacian eigenvalues λ_i's satisfy (see [22])

$$
|\lambda_i - 1| \leq \frac{(3 + o_n(1))}{\sqrt{np}}. \tag{12}
$$

Apply Lemma 2 with $\epsilon = \frac{(3+o_n(1))}{\sqrt{np}}$, and note that $G(n,p)$ is almost-regular. We get the following theorem.

Theorem 5. *For* $p(1-p) \gg \frac{\log^4 n}{n}$, *almost surely for all pairs of vertices* (u,v), *the DSD* L_q-*distance of* $G(n,p)$ *satisfies*

$$DSD_q(u,v) = 2^{1/q} \pm O\left(\frac{1}{\sqrt{np}}\right) \quad if \ q \geq 2,$$

$$DSD_q(u,v) = 2^{1/q} \pm O\left(\frac{n^{\frac{1}{q}-\frac{1}{2}}}{\sqrt{np}}\right) \quad if \ 1 \leq q < 2.$$

Now we consider the random graphs with given expected degree sequence $G(w_1,\ldots,w_n)$ (see [2], [7], [8], [9], [20]). It is defined as follows:

1. Each vertex i (for $1 \leq i \leq n$) is associated with a given positive weight w_i.
2. Let $\rho = \frac{1}{\sum_{i=1}^{n} w_i}$. For each pair of vertices (i,j), ij is added as an edge with probability $w_i w_j \rho$ independently. (i and j may be equal so loops are allowed. Assume $w_i w_j \rho \leq 1$ for i,j.)

Let w_{min} be the minimum weight. There are many references on the concentration of the eigenvalues of $G(w_1,\ldots,w_n)$ (see [10], [11], [12], [14], [22]). The version used here is in [22].

1. For each vertex i, the expected degree of i is w_i.
2. Almost surely for all i with $w_i \gg \log n$, then the degree $d_i = (1+o(1))w_i$.
3. If $w_{min} \gg \log^4 n$, all non-zero Laplacian eigenvalues λ_i (for $1 \leq i \leq n-1$),

$$|1-\lambda_i| \leq \frac{3+o_n(1)}{\sqrt{w_{min}}}. \tag{13}$$

Theorem 6. *Suppose* $w_{min} \gg \log^4 n$, *almost surely for all pairs of vertices* (u,v), *the DSD* L_q-*distance of* $G(w_1,\ldots,w_n)$ *satisfies*

$$DSD_q(u,v) = 2^{1/q} \pm O\left(\frac{1}{\sqrt{w_{min}}}\sqrt{\frac{w_{max}}{w_u} + \frac{w_{max}}{w_v}}\right) \quad if \ q \geq 2,$$

$$DSD_q(u,v) = 2^{1/q} \pm O\left(\frac{n^{\frac{1}{q}-\frac{1}{2}}}{\sqrt{w_{min}}}\sqrt{\frac{w_{max}}{w_u} + \frac{w_{max}}{w_v}}\right) \quad if \ 1 \leq q < 2.$$

6   Examples of Biological Networks

In this section, we will examine the distribution of the DSD distances for some biological networks. The set of graphs analyzed in this section include three graphs of brain data from the Open Connectome Project [25] and two more graphs built from the *S. cerevisiae* PPI network and *S. pombe* PPI network used in [3]. Figure 1 and 2 serves as a visual representation of one of the two brain

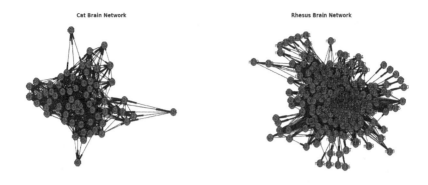

Fig. 1. The brain networks: (a), a Cat; (b): a Rhesus Monkey

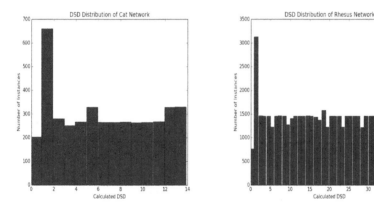

Fig. 2. The distribution of the DSD L_1-distances of brain networks: (a), a Cat; (b): a Rhesus Monkey

data graphs: the graph of a cat and the graph of a Rhesus monkey. The network of the cat brain has 65 nodes and 1139 edges while the network of rhesus monkey brain has 242 nodes and 4090 edges.

Each node in the Rhesus graph represents a region in the cerebral cortex originally analyzed in [18]. Each edge represents axonal connectivity between regions and there is no distinction between strong and weak connections in this graph [18]. The Cat data-set follows a similar pattern where each node represents a region of the brain and each edge represents connections between them. The Cat data-set represents 18 visual regions, 10 auditory regions, 18 somatomotor regions, and 19 frontolimbic regions[23].

For each network above, we calculated all-pair DSD L_1-distances. Divide the possible values into many small intervals and compute the number of pairs

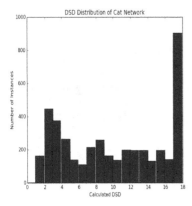
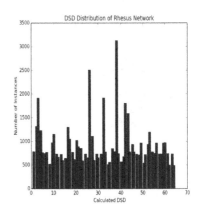

Fig. 3. The distribution of the DSD L_2-distances of brain networks: (a), a Cat; (b): a Rhesus Monkey

falling into each interval. The results are shown in Figure 1. The patterns are quite surprising to us.

Both graphs has a small interval consisting of many pairs while other values are more or less uniformly distributed. We think, that phenomenon might be caused by the clustering of a dense core. The two graphs have many branches sticking out. Since we are using L_1-distance, it doesn't matter the directions of these branches sticking out when they are embedded into $\mathbb{R}^n$ using Green's function.

When we change L_1-distance to L_2-distance, the pattern should be broken. This is confirmed in Figure 3. The actual distributions are mysterious to us.

References

1. Baroni, S., Giannozzi, P., Testa, A.: Greens-function approach to linear response in solids. Physical Review Letters **58**(18), 1861–1864 (1987)
2. Bhamidi, S., van der Hofstad, R.W., van Leeuwaarden, J.S.H.: Scaling limits for critical inhomogeneous random graphs with finite third moments. Electronic Journal of Probability **15**(54), 1682–1702 (2010)
3. Cao, M., Zhang, H., Park, J., Daniels, N.M., Crovella, M.E., Cowen, L.J., Hescott, B.: Going the distance for protein function prediction: a new distance metric for protein interaction networks. PLoS ONE **8**(10), e76339 (2013)
4. Chin, S.P., Reilly, E., Lu, L.: Finding structures in large-scale graphs. SPIE Defense, Security, and Sensing. International Society for Optics and Photonics (2012)
5. Chung, F., Yau, S.-T.: Discrete Green's functions. J. Combinatorial Theory (A) **91**, 191–214 (2000)
6. Chung, F.: Spectral graph theory. AMS Publications (1997)
7. Chung, F., Lu, L.: Connected components in a random graph with given degree sequences. Annals of Combinatorics **6**, 125–145 (2002)

8. Chung, F., Lu, L.: The average distances in random graphs with given expected degrees. Proc. Natl. Acad. Sci. **99**, 15879–15882 (2002)
9. Chung, F., Lu, L.: Complex graphs and networks. CBMS Regional Conference Series in Mathematics, vol. 107, 264+vii p. (2006). ISBN-10: 0–8218-3657-9, ISBN-13: 978-0-8218-3657-6
10. Chung, F., Lu, L., Vu, V.H.: Eigenvalues of random power law graphs. Ann. Comb. **7**, 21–33 (2003)
11. Chung, F., Lu, L., Vu, V.H.: Spectra of random graphs with given expected degrees. Proc. Natl. Acad. Sci. USA **100**(11), 6313–6318 (2003)
12. Chung, F., Radcliffe, M.: On the spectra of general random graphs. Electron. J. Combin. **18**(1), P215 (2011)
13. Coja-Oghlan, A.: On the Laplacian eigenvalues of $G(n,p)$. Combin. Probab. Comput. **16**(6), 923–946 (2007)
14. Coja-Oghlan, A., Lanka, A.: The spectral gap of random graphs with given expected degrees. Electron. J. Combin. **16**, R138 (2009)
15. Davies, E.B., Kernels, H.: Spectral Theory, vol. 92. Cambridge University Press (1990)
16. Duffy, D.G.: Green's Functions with Applications. CRC Press (2010)
17. Green, G.: An Essay on the Application of Mathematical Analysis to the Theories of Electricity and Magnetism. Nottingham (1828)
18. Harriger, L., van den Heuvel, M.P., Sporns, O.: Rich club organization of macaque cerebral cortex and its role in network communication. PloS One **7**(9), e46497 (2012)
19. Hedin, L.: New method for calculating the one-particle Green's function with application to the electron-gas problem. Physical Review **139**(3A), 796–823 (1965)
20. van der Hofstad, R.W.: Critical behavior in inhomogeneous random graphs. Random Structures and Algorithms **42**(4), 480–508 (2013)
21. Krivelevich, M., Sudakov, B.: The largest eigenvalue of sparse random graphs. Combin. Probab. Comput. **12**, 61–72 (2003)
22. Lu, L., Peng, X.: Spectra of edge-independent random graphs. Electronic Journal of Combinatorics **20**(4), P27 (2013)
23. de Reus, M.A., van den Heuvel, M.P.: Rich club organization and intermodule communication in the cat connectome. The Journal of Neuroscience **33**(32), 12929–12939 (2013)
24. Stakgold, I., Holst, M.J.: Green's Functions and Boundary Value Problems. John Wiley & Sons (2011)
25. Open Connectome Project, Web (June 25, 2014). http://www.openconnectomeproject.org

Relational Topic Factorization for Link Prediction in Document Networks

Wei Zhang[1,2]([⊠]), Jiankou Li[1,2], and Xi Yong[1,2]

[1] State Key Laboratory of Computer Science, Institute of Software Chinese Academy of Sciences, P.O. Box 8718, Beijing 100190, People's Republic of China
zhangw@ios.ac.cn
[2] School of Information Science and Engineering, University of Chinese Academy of Sciences, Beijing, People's Republic of China
http://www.springer.com/lncs

Abstract. Link prediction is one of the fundamental problems in complex networks. In this paper, we focus on link prediction in document networks, in which nodes are text documents. We propose the relational topic factorization model (RTF), a model that combines topic models and matrix factorization. We also develop an efficient Monte Carlo EM algorithm for learning the parameters. Empirical results show that our model outperforms other state-of-the-art ones, and can give better understanding of the documents.

Keywords: Link prediction · Matrix factorization · Latent Dirichlet allocation

1 Introduction

In the real world, many kinds of data can be represented by complex networks, e.g. social networks, citation networks, communication networks and protein interaction networks. The study on complex networks has attracted researchers from many different fields. Among the studies, one of the fundamental problems is *link prediction* that aims at identifying structural patterns of the networks and predict missing links among the nodes.

Earlier studies on link prediction are mainly based on analyzing link structure of networks [9,14,16]. These models can be used in all link prediction tasks since they ignore the node type. Recently, more complex models are proposed by taking node attributes into account [3,11,13].

In this paper, we focus on link prediction in document networks, in which nodes are text documents (e.g., citation networks and web graphs). The model also applies to other discrete data types (e.g., social networks in which each user is marked with tags).

Clearly, a full model for both links and node attributes can provide more precise predictions, and make predictions on new documents that using only their

© Springer International Publishing Switzerland 2014
A. Bonato et al. (Eds.): WAW 2014, LNCS 8882, pp. 96–107, 2014.
DOI: 10.1007/978-3-319-13123-8_8

content. This is intuitive, since link patterns and node attributes are closely inter-related: documents with similar content are more likely to be linked together. Meanwhile, the model should also be flexible enough to allow link patterns diverge from node attributes. For instance, the reason why a web page points to another web page may be simply because that the latter is popular and there is no content similarity between them. These link patterns cannot be interpreted by node attributes and have to be learned from the network structure.

However, previous models cannot well balance the influence of links and node attributes. In this paper, we propose a probabilistic generative model for document networks, which combines topic models and matrix factorization. We use topic models to extract topic structures of documents, and use a coupled matrix factorization to learn link patterns between documents. We express relationships between topic structures and link patterns by regression models that connect latent variables of topic models and matrix factorization. The flexibility is ensured by the smooth connections and well controlled by the parameters. Furthermore, we also distinguish between outgoing links and ingoing links. For a document, patterns of these two kinds of links may be different. For instance, a mathematical article mainly citing other mathematical articles may be cited by diverse domains such as data mining or biology. We evaluate our model on scientific articles from data mining domain. Empirical results showed that our model outperforms other state-of-the-art models on link prediction tasks. Moreover, our model can reveal deep properties of the documents.

The remainder of the paper is organized as follows. In section 2, we give the background, and review two basic models. In section 3, we present our model and develop a Monte Carlo EM algorithm to learn the parameters. In section 4, we evaluate our model via experimental study. Finally, we conclude our work in section 5.

2   Related Work

Earlier studies on link prediction are mainly based on analyzing network structures [9,14,16]. Many of them are latent factor models, e.g., mixed membership stochastic blockmodels (MMSB) [1], matrix factorization (MF) [12], and other models [5,6,8].

Recently, joint models for both links and node attributes are also proposed. For discrete node attributes such as text, extensions that incorporating topic models for document networks have also been proposed [3,4,11,13,19]. As another related learning task, community detection is also studied by considering node attributes [2,10,17,18]. As link prediction has intimate connection with recommendation tasks, recently, researchers also employ MF for link prediction tasks [12]. Our work is also inspired by models for recommender systems [7,15].

Among these models, our model is closely related to RTM [3] and CTR [15]. Compared with RTM, our model use matrix factorization to model link patterns instead of using direct response functions. Compared with CTR, we are considering different learning tasks in our model (CTR is a model for recommender systems), and we also develop a different learning algorithm.

3   Proposed Model

In this section, we describe our model–the relational topic factorization (RTF). RTF combines topic modeling with matrix factorization. We use latent Dirichlet allocation to model the content of documents, and use a coupled matrix factorization to model the links between documents. We first present the definition of RTF, and then we develop an efficient Monte Carlo EM algorithm for learning the parameters.

3.1   Relational Topic Factorization

RTF is a generative model for documents and links between them. In RTF, each document is generated from its corresponding topic proportions. Links between documents are then generated according to their topic proportions. This is reasonable, since documents with similar topics are more likely to be linked together.

Let z_{dn} be the topic assignment of the nth word in document d, we can summarize the topics of document d as the average topic assignments $\bar{\mathbf{z}}_d$. For a document pair (d, d'), we wish to describe the generating process of the link between them. One approach is to directly define a response function for the link indicator variable $y_{dd'}$,

$$y_{dd'} \sim \psi(\cdot | \bar{\mathbf{z}}_d, \bar{\mathbf{z}}_{d'}). \tag{1}$$

This is the approach adopted by relational topic modeling (RTM) [3]. The response function is usually defined as the sigmoid function $\psi_\sigma(y_{dd'} = 1) = \sigma(\eta^T(\bar{\mathbf{z}}_d \circ \bar{\mathbf{z}}_{d'}) + \nu)$ or the exponential function $\psi_t(y_{dd'} = 1) = exp(\eta^T(\bar{\mathbf{z}}_d \circ \bar{\mathbf{z}}_{d'}) + \nu)$, where $\circ$ denotes the Hadamard (element-wise) product, and $\sigma(\cdot)$ is the sigmoid function.

However, this approach limits the descriptive ability of the model, since links are based heavily on contents. In the real world, many links cannot be well predicted from contents. For instance, the reason why a web page points to another may be simply that the latter is popular, and there is no content similarity between them.

In RTF, we introduce two latent vectors u_d and v_d for each document d. For each document pair (d, d'), we define the response function as

$$y_{dd'} \sim \mathcal{N}(u_d^T v_{d'}, c_{dd'}^{-1})$$

Let matrices $U = (u_1, \cdots, u_d)$ and $V = (v_1, \cdots, v_d)$, this is equivalent to the assumption that the adjacency matrix of the network can be factorized into the product of two low-rank matrices U^T and V.

To describe the relationship between links and document contents, we further specify that the prior of u_d and v_d is the topic proportions θ_d of document d

$$\begin{aligned} u_d &\sim \mathcal{N}(\theta_d, \lambda_u^{-1}\mathbf{I}) \\ v_d &\sim \mathcal{N}(\theta_d, \lambda_v^{-1}\mathbf{I}) \end{aligned} \tag{2}$$

This can be considered as a soft constraint between links and contents. In fact, if we integrate out u_d and v_d, we obtain the following response function based purely on topic proportions

$$
\begin{aligned}
y_{dd'}|\theta_d, \theta_{d'} \sim \int &\mathcal{N}(u_d|\theta_d, \lambda_u^{-1}\mathbf{I})\mathcal{N}(v_{d'}|\theta_{d'}, \lambda_v^{-1}\mathbf{I}) \\
&\mathcal{N}(y_{dd'}|u_d^T v_{d'}, c_{dd'}^{-1})du_d dv_{d'}
\end{aligned}
\tag{3}
$$

where λ_u, λ_v and $c_{dd'}$ are precision parameters. This is a complex function that has no close form, by which we gain the ability to describe more complex link patterns.

Based on previous discussions, we define the generative process of RTF as

1. For each topic k,
   draw word distribution $\phi_k \sim Dir(\beta)$.
2. For each document d,
   (a) Draw topic proportions $\theta_d \sim Dir(\alpha)$.
   (b) For each word w_{dn},
       i. Draw topic assignment $z_{dn} \sim Mult(\theta_d)$
       ii. Draw word $w_{dn} \sim Mult(\phi_{z_{dn}})$
   (c) Draw latent factors $u_d \sim \mathcal{N}(\theta_d, \lambda_u^{-1}\mathbf{I})$.
   (d) Draw latent factors $v_d \sim \mathcal{N}(\theta_d, \lambda_v^{-1}\mathbf{I})$.
3. For each pair of documents (d, d'), draw binary link indicator

$$
y_{dd'} \sim \mathcal{N}(u_d^T v_{d'}, c_{dd'}^{-1})
$$

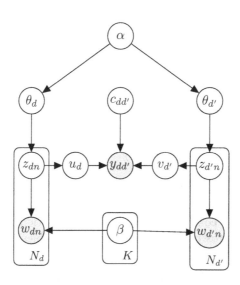

Fig. 1. Graphical representation of RTF

Figure 1 shows the graphical representation of RTF. In RTF, each document d has three representations: topic proportions θ_d that represents d's content; outgoing latent factors u_d that represents d's citation pattern to other documents; ingoing latent factors v_d that represents citations patterns of other documents to d. Relationships between these three representations reveal interesting properties of the documents.

3.2   Learning the Parameters

In this subsection, we develop a Monte Carlo EM algorithm to learn the parameters of RTF. In E-step, we use collapsed Gibbs sampling to approximate the expectations of latent variables; in M-step, we use alternating least squares to optimize the parameters.

The full joint distribution of RTF is

$$p(\mathbf{w}, \mathbf{y}, \mathbf{z}, U, V, \Theta, \Phi | \alpha, \beta, C)$$

$$= \prod_k p(\phi_k | \beta) \prod_d p(\theta_d | \alpha) = \prod_{dn} p(w_{dn} | \phi_{z_{dn}}) \prod_{dn} p(z_{dn} | \theta_d) \tag{4}$$

$$= \prod_d p(u_d | z_d) \prod_d p(v_d | z_d) = \prod_{dd'} p(y_{dd'} | u_d, u_{d'}, c_{dd'})$$

where $\{\mathbf{w}, \mathbf{y}\}$ is the set of observations, $\{\mathbf{z}, \Theta, \Phi\}$ is the set of latent variables, and $\{U, V\}$ is the set of parameters.

E-step

In E-step, our goal is to compute the posterior distribution of latent variables $\{\mathbf{z}, \Theta, \Phi\}$, given the parameters $\{U, V\}$:

$$p(\mathbf{z}, \Theta, \Phi | U, V) \tag{5}$$

where observations $\mathbf{y}, \mathbf{w}$, and hyperparameters α, β, C are omitted for succinctness. As in LDA, we integrate out Θ and Φ to obtain the collapsed joint distribution:

$$p(\mathbf{w}, \mathbf{z}, U, V)$$

$$= \prod_d \left[\frac{\Gamma(K\alpha)}{\Gamma(N_d + K\alpha)} \prod_k \frac{\Gamma(N_{dk} + \alpha)}{\Gamma\alpha} \right] \prod_k \left[\frac{\Gamma(W\beta)}{\Gamma(N_k + W\beta)} \prod_w \frac{\Gamma(N_{kw} + \beta)}{\Gamma\beta} \right] \tag{6}$$

$$\prod_d \mathcal{N}(u_d | z_d, \lambda_u I) \prod_{d'} \mathcal{N}(v_{d'} | z_{d'}, \lambda_v I)$$

where K is the number of topics, W is the vocabulary size, and $N_{..}$ are counting variables.

We cannot directly compute the posterior distribution $p(\mathbf{z} | U, V)$, so we use Gibbs sampling to approximate it. With all other variables fixed, the conditional distribution for sampling z_{dn} is given by:

$$p(z_{dn} = k | z^{\neg dn}, u_d, v_d, w_{dn} = w)$$

$$\propto \left[(N_{dk}^{\neg dn} + \alpha) \frac{N_{kw}^{\neg dn} + \beta}{N_k^{\neg dn} + W\beta} \right] \mathcal{N}(u_d | z_d, \lambda_u I) \mathcal{N}(v_{d'} | z_{d'}, \lambda_v I) \tag{7}$$

The result is quite intuitive: it is the conditional distribution from LDA, multiplied by the likelihood term of the latent factors. We initialize z_d variables and then iteratively sample each z_{dn} from the corresponding conditional distribution. After sufficiently many iterations, we collect sufficient statistics of z_d's, which will be used in M-step for optimizing the parameters. It turns out that we only need to compute the sample mean of the each z_{dn} and record the average value $\mathbb{E}[\bar{z}_d]$ for each document.

M-step

In M-step, our goal is to optimize parameters U, V by maximizing the expectations of the complete log-likelihood.

$$argmax_{U,V} \sum_{\mathbf{z}} p(\mathbf{z}|U^{old}, V^{old}) \ln p(\mathbf{z}, \mathbf{y}|U, V) \tag{8}$$

We separate items containing U, V, and obtain the following optimization problem:

$$argmin_{U,V} \frac{\lambda_u}{2} \sum_d (u_d - \mathbb{E}[\bar{z}_d])^T (u_d - \mathbb{E}[\bar{z}_d]) + \frac{\lambda_v}{2} \sum_{d'} (v_{d'} - \mathbb{E}[\bar{z}_{d'}])^T (v_{d'} - \mathbb{E}[\bar{z}_{d'}])$$
$$+ \sum_{dd'} \frac{c_{dd'}}{2} (y_{dd'} - u_d^T v_{d'})^2 \tag{9}$$

We use alternating least squares to optimize U and V. With V fixed, the problem becomes quadratic with respect to U. We compute gradients with respect to u_d's, and set them to zero to obtain the following updating equation:

$$u_d = (V C_d V^T + \lambda_u I_K)^{-1} (V C_d Y_d + \lambda_u \mathbb{E}[\bar{z}_d]) \tag{10}$$

where C_d is a diagonal matrix with $c_{dd'}, d' = 1 \cdots, D$ as its diagonal elements and $Y_d = (y_{dd'})_{d'=1}^D$. Similarly, we can optimize $v_{d'}$ by the following equation:

$$v_{d'} = (U C_{d'} U^T + \lambda_v I_K)^{-1} (U C_{d'} Y_{d'} + \lambda_v \mathbb{E}[\bar{z}_{d'}]) \tag{11}$$

where $C_{d'}$ and $Y_{d'}$ are similarly defined. For further details, the reader is referred to [7].

With these updating equations, we alternatively optimize U and V until convergence conditions are achieved.

Prediction

We consider two prediction tasks in this paper. The first task is in-matrix prediction, which means predicting links based on contents and previous links. In general, the prediction is estimated by,

$$\mathbb{E}[y_{dd'}|D] \approx \mathbb{E}[u_d|D]^T \mathbb{E}[v_{d'}|D] \tag{12}$$

We approximate these expectations using the point estimations of $u_d, v_{d'}$,

$$y_{dd'}^* = (u_d^*)^T v_{d'}^* \tag{13}$$

The second task is out-of-matrix prediction, which means predicting links for a new document based purely on its content. For a new document d, we cannot infer its latent factors, but we can still infer its topic proportions θ_d by Gibbs sampling.

There are three possibilities. For links from a new document d_{new} to an old document d_{old}, we approximate the prediction as:

$$y^*_{d_{new}d_{old}} = (\mathbb{E}[\theta_{d_{new}}])^T v^*_{d_{old}} \tag{14}$$

For links from an old document d_{old} to a new document d_{new}, we approximate the prediction as:

$$y^*_{d_{old}d_{new}} = (u^*_{d_{old}})^T \mathbb{E}[\theta_{d_{new}}] \tag{15}$$

For links between two new documents d and d', we approximate the prediction as:

$$y^*_{dd'} = \mathbb{E}[\theta_d]^T \mathbb{E}[\theta_{d'}] \tag{16}$$

4   Empirical Results

4.1   Dataset

The data we used are scientific articles from data mining domain, which was collected from Microsoft Academy Search Engine[1]. For each article, we have its title, abstract, authors, and references. We removed articles that do not have abstracts to obtain a dataset that contains 56280 articles and 219293 citation links.

For each article, we concatenated its title and abstract. After removing stop words, we use tf-idf to choose the top 8,000 distinct words as the vocabulary. Finally, we obtain a corpus of 4.3M words.

4.2   Evaluation Metrics

In this paper, we focus on link prediction problems. Our task is to predict links for a document based on its content and(or) its previous links. Precision and recall are two alternative metrics. In our experiment, precision cannot be correctly computed. This is because that missing links are uncertain–the author may be not aware of other documents. Hence, we choose recall as our evaluation metric. In our model, we distinguish outgoing and ingoing links. But for evaluation and comparisons, we treat links as undirected ones. For each document d, we suggest M links for it based on the average predictions for outgoing and ingoing links on other documents. We then define the *recall@M* as:

$$recall@M = \frac{\text{number of true links in all predictions}}{\text{total number of true links}} \tag{17}$$

Note that we calculate only once for each document pair (d, d').

4.3   In-matrix Prediction

We use 5-fold cross validation to evaluate the performance. For every document that has as least 5 outgoing links, we randomly split the links into 5 even folds. At each iteration, we consider one of them as test set, with the others as training set.

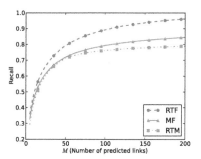
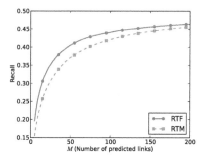

Fig. 2. In-matrix prediction   **Fig. 3.** Out-of-matrix prediction

We compare our model (RTF) with relational topic models (RTM) and matrix factorization (MF). For all models, we set $K = 50$. For MF, we set $\lambda_u = \lambda_v = 0.01$; for RTF, we set $\lambda_u = \lambda_v = 1$; For RTM, we use the exponential response function.

Figure 2 shows the comparison of three models. We observe that RTF outperforms other two models, and MF also outperforms RTM in this task. This is because that in RTM, response functions are restricted, and the predictions are based heavily on content. In contrast, RTF can better balance the effect of content and links, and hence make more precise predictions.

Impact of Parameters

In this subsection, we investigate the impact of parameters λ_u and λ_v. These precision parameters control how much we allow that the latent factors of documents can diverge from their topic proportions: large λ_u means that documents tend to cite other documents that have similar content; large λ_v means that documents tend to be cited by other documents with similar content.

We vary λ_u and λ_v ranging from 0.001 to 1000, fit the model with respect to different parameter pairs, and compute the Recall@50 for each pair. From Figure 4, we observe that the effect of λ_u and λ_v is symmetrical, and the performance reach the optimum when $\lambda_u \cdot \lambda_v \leq 1$. Therefore, we set $\lambda_u = \lambda_v = 1$ in our experiment for symmetry.

4.4   Out-of-Matrix Prediction

We again use 5-fold cross validation. We randomly group all documents into 5 folds. For documents in each folder, we pick out all links on them

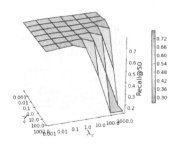

Fig. 4. Impact of parameters λ_u and λ_v

(including outgoing and ingoing links) to build the test set, and leave the remaining links as training set. We then iteratively fit the models to each training set, and evaluate the performance on the corresponding test set. MF cannot perform out-of-matrix prediction, so we only compare RTF with RTM. From Figure 3, we observe that RTF also outperforms RTM on this task.

4.5   Relationship with Document Properties

In this subsection, we study how the performance and parameters of our model vary as functions of object properties: number of outgoing links and number of ingoing links.

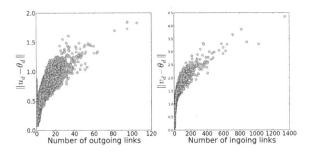

Fig. 5. The left plot shows how the distances between θ_d and u_d vary as functions of the number of outgoing links of document d; the right plot shows how the distances between θ_d and v_d vary as functions of the number of ingoing links of document d

Recall that in RTF, each document d has three representations: topic proportions θ_d, outgoing latent factors u_d, and ingoing latent factors v_d. From the left plot of Figure 5, we observe that as a document cites more other documents,

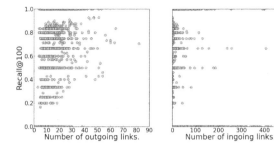

Fig. 6. These scatter plots show how the recall varies as functions of the number of outgoing and ingoing links of documents

the distances between u_d and θ_d tend to increase. This is reasonable, because that documents citing more others are likely to cite more diverse documents.

Similar results can be observed in the right plot of Figure 5, which means that popular documents tend to be cited by more diverse documents.

We then study the relationship between performance and document properties. In Figure 6, we plot recall@100 as a function of the number of articles a document cites. We observe an upward trend from both sides of plots, which mean that both outgoing and ingoing links help to make more precise predictions.

4.6   Examining Topic Spaces

In this subsection, we examine topic spaces learned by RTF. For each document d, we can interpret its topics from three point of views corresponding to its three representations θ_d, u_d and v_d. θ_d represents topic proportions based on d's content; u_d represents average topic proportions of documents that are cited by d; v_d represents average topic proportions of documents that cite d.

We rank entries of each representation vector, and choose top topics for two example articles. The first article is "Data Mining: Concepts and Techniques", which is one of the top cited articles in the field of data mining. From Table 1, we observe that topics according to its content (θ_d) are general topics about data mining, and topics according to u_d and v_d are diverse topics of specific subdomains. This is because it is a survey article that cites many articles of different subdomains, and it is also cited by other articles from many different subdomains.

The second article is "Relational Topic Models for Document Networks". From Table 2, we observe that topics ranked by three representations are similar. This is because the article is mainly about topic models applying to document networks, and both articles citing it and articles it cites are of similar themes.

Table 1. Top topics of article "Data Mining: Concepts and Techniques", ranked according to θ_d, u_d, and v_d

Top topics ranked by θ_d
1. data, mine, mining, representation, learning, field, survey, task
2. pattern, database, algorithm, online, experiment, outperform, sequential
3. discovery, dataset, emerge, pattern, hidden, extract, representation
Top topics ranked by u_d
1. space, dimension, subspace, high-dimension, neighborhood, vector
2. cluster, algorithm, data, k-mean, clusters, category, clustering, hierachy
3. feature, select, subset, features, base, selection, filter, candidate
Top topics ranked by v_d
1. scientific, challenge, technic, science, research, technology, workshop
2. Wikipedia, coverage, edit, book, supply, article, title, gender
3. track, face, background, target, motion, moment, camera, obstacle

Table 2. Top topics of article "Relational Topic Models for Document Networks", ranked according to θ_d, u_d, and v_d

Top topics ranked by θ_d
1. document, collection, text, documents, topic, paper, corpus, representation
2. network, communication, networks, connect, node, link, complex, topology
3. model, mixture, latent, hidden, infer, probabilistic, generative, factor
Top topics ranked by u_d
1. document, collection, text, documents, topic, paper, corpus, representation
2. network, communication, networks, connect, node, link, complex, topology
3. model, mixture, latent, hidden, infer, probabilistic, generative, factor
Top topics ranked by v_d
1. topic, author, paper, citation, research, publication, article, academy
2. model, mixture, latent, hidden, infer, probabilistic, generative, factor
3. network, communication, networks, connect, node, link, complex, topology

5 Conclusions and Future Work

In this paper, we proposed a probabilistic generative model for document networks. The model (RTF) combines latent Dirichlet allocation and matrix factorization. We evaluate our model on link prediction tasks. Empirical results demonstrate that our model outperforms other state-of-the-art ones. Moreover, our model can reveal deep properties of documents. For future work, we are interested in parallelizing our algorithm and examining it on large-scale datasets.

References

1. Airoldi, E.M., Blei, D.M., Fienberg, S.E., Xing, E.P.: Mixed Membership Stochastic Blockmodels. Journal of Machine Learning Research **9**, 33–40 (2008)
2. Balasubramanyan, R., Cohen, W.W. Block-lda: Jointly modeling entity-annotated text and entity-entity links. In: SDM, vol. 11, pp. 450–461. SIAM (2011)

3. Chang, J., Blei, D.M.: Relational topic models for document networks. In: International Conference on Artificial Intelligence and Statistics, pp. 81–88 (2009)
4. Chen, N., Zhu, J., Xia, F., Zhang, B.: Generalized relational topic models with data augmentation. In: Proceedings of the Twenty-Third International Joint Conference on Artificial Intelligence, pp. 1273–1279. AAAI Press (2013)
5. Hoff, P.D., Raftery, A.E., Handcock, M.S.: Latent space approaches to social network analysis. Journal of the American Statistical Association **97**(460), 1090–1098 (2002)
6. Hofman, J.M., Wiggins, C.H.: Bayesian approach to network modularity. Physical Review Letters **100**(25), 258701 (2008)
7. Hu, Y., Koren, Y., Volinsky, C.: Collaborative filtering for implicit feedback datasets. In: 2008 Eighth IEEE International Conference on Data Mining, ICDM 2008, pp. 263–272. IEEE (2008)
8. Kemp, C., Griffiths, T.L., Tenenbaum, J.B.: Discovering latent classes in relational data
9. Liben-Nowell, D., Kleinberg, J.M.: The link prediction problem for social networks. In: International Conference on Information and Knowledge Management, pp. 556–559 (2003)
10. Liu, Y., Niculescu-Mizil, A., Gryc, W.: Topic-link lda: joint models of topic and author community. In: Proceedings of the 26th Annual International Conference on Machine Learning, pp. 665–672. ACM (2009)
11. Mei, Q., Cai, D., Zhang, D., Zhai, C.: Topic modeling with network regularization. In: Proceedings of the 17th International Conference on World Wide Web, pp. 101–110. ACM (2008)
12. Menon, A.K., Elkan, C.: Link prediction via matrix factorization. In: Gunopulos, D., Hofmann, T., Malerba, D., Vazirgiannis, M. (eds.) ECML PKDD 2011, Part II. LNCS, vol. 6912, pp. 437–452. Springer, Heidelberg (2011)
13. Nallapati, R., Cohen, W.W.: Link-plsa-lda: A new unsupervised model for topics and influence of blogs. In: ICWSM (2008)
14. Newman, M.E.J.: The Structure and Function of Complex Networks. Siam Review 45 (2003)
15. Wang, C., Blei, D.M.: Collaborative topic modeling for recommending scientific articles. In: Proceedings of the 17th ACM SIGKDD International Conference on Knowledge Discovery and Data Mining, pp. 448–456. ACM (2011)
16. Wasserman, S., Pattison, P.: Logit models and logistic regressions for social networks: I. an introduction to markov graphs andp. Psychometrika **61**(3), 401–425 (1996)
17. Xu, Z., Ke, Y., Wang, Y., Cheng, H., Cheng, J.: A model-based approach to attributed graph clustering. In: Proceedings of the 2012 ACM SIGMOD International Conference on Management of Data, pp. 505–516. ACM (2012)
18. Yang, J., McAuley, J., Leskovec, J.: Community detection in networks with node attributes. In: 2013 IEEE 13th International Conference on Data Mining (ICDM), pp. 1151–1156. IEEE (2013)
19. Zhu, Y., Yan, X., Getoor, L., Moore, C.: Scalable text and link analysis with mixed-topic link models. In: Proceedings of the 19th ACM SIGKDD International Conference on Knowledge Discovery and Data Mining, pp. 473–481. ACM (2013)

Firefighting as a Game

Carme Àlvarez, Maria J. Blesa, and Hendrik Molter[✉]

ALBCOM Research Group, Computer Science Department,
Universitat Politècnica de Catalunya, BarcelonaTech, 08034 Barcelona, Spain
{alvarez,mjblesa}@cs.upc.edu, hendrik.molter@gmail.com

Abstract. The Firefighter Problem was proposed in 1995 [16] as a deterministic discrete-time model for the spread (and containment) of a fire. Its applications reach from real fires to the spreading of diseases and the containment of floods. Furthermore, it can be used to model the spread of computer viruses or viral marketing in communication networks.

In this work, we study the problem from a game-theoretical perspective. Such a context seems very appropriate when applied to large networks, where entities may act and make decisions based on their own interests, without global coordination.

We model the Firefighter Problem as a strategic game where there is one player for each time step who decides where to place the firefighters. We show that the Price of Anarchy is linear in the general case, but at most 2 for trees. We prove that the quality of the equilibria improves when allowing coalitional cooperation among players. In general, we have that the Price of Anarchy is in $\Theta(\frac{n}{k})$ where k is the coalition size. Furthermore, we show that there are topologies which have a constant Price of Anarchy even when constant sized coalitions are considered.

Keywords: Firefighter problem · Spreading models for networks · Algorithmic game theory · Nash equilibria · Price of anarchy · Coalitions

1 Introduction

The Firefighter Problem was introduced by Hartnell [16] as a deterministic discrete-time model for the spread and containment of fire. Since then, it has been subject to a wide variety of research for modeling spreading and containment phenomena like diseases, floods, ideas in social networks and viral marketing.

The Firefighter Problem takes place on an undirected finite graph $G = (V, E)$, where initially fire breaks out at f nodes. In each subsequent time-step, two actions occur: A certain number b of firefighters are placed on non-burning nodes, permanently protecting them from the fire. Then the fire spreads to all non-defended neighbors of the vertices on fire. Since the graph is finite, at some

This work was supported by grants TIN2013-46181-C2-1-R, TIN2012-37930 and grant TIN2007-66523 of the Spanish Government, and project 2014-SGR1034 of the Generalitat de Catalunya.

© Springer International Publishing Switzerland 2014
A. Bonato et al. (Eds.): WAW 2014, LNCS 8882, pp. 108–119, 2014.
DOI: 10.1007/978-3-319-13123-8_9

point each vertex is either on fire or saved. Then the process finishes, because the fire cannot spread any further. There are several different objectives for the problem. Typically, the goal is to save the maximum possible number of nodes. Other objectives include minimizing the number of firefighters (or time-steps) until the spreading stops, or determining whether all vertices in a specified collection can be prevented from burning.

Most research on the Firefighter Problem (also the work in this paper) considers the case $f = b = 1$, which already leads to hard problems. The problem was proved NP-hard for bipartite graphs [20], graphs with degree three [11], cubic graphs [19] and unit disk graphs [14]. However, the problem is polynomial-time solvable for various well-known graph classes, including interval graphs, split graphs, permutation graphs, caterpillars, and P_k-free graphs for fixed k [12,14, 15,20]. Furthermore, the problem is $(1 - 1/e)$-approximable on general trees [6], 1.3997-approximable for trees where vertices have at most three children [18], and it is NP-hard to approximate within $n^{(1-\varepsilon)}$ for any $\varepsilon > 0$ [3]. Later results on approximability for several variants of the problem can be found in [3,5,8].

Recently, the scientific community has focused on the study of the parameterized complexity of the problem. It was shown to be fixed parameter-tractable w.r.t. combined parameter "pathwidth" and "maximum degree" [7]. Other important results can be found in [4,9]. For other variants of the Firefighter Problem see [10,12,21].

In this work, we study the Firefighter Problem from a game-theoretical perspective. Instead of global coordination algorithms, we define a game where the players decide which nodes to protect. Player i chooses where to place the firefighters at time-step i, independently from the other players (one shot game). Since we consider the case of $b = 1$, every player can protect at most one node in his corresponding turn. We can consider different payoffs for the players, the most natural seems to save as many nodes as possible. At each time-step, the fire spreads automatically as described in the original problem.

To the best of our knowledge, the only existing game-theoretical models to similar problems are those referred to as the vaccination problem [3,13], the spreading of rumors [25] and competitive diffusion [1,22–24]. Those models however focus on information spreading on social networks, and thus take into account other inherent aspects of those scenarios, like preferences, reputation, popularity and other personal traits of the users, and relevance or truthfulness of the information. Our proposal is well-suited to model fighting against spreading phenomena in large networks, where the protection strategy for each time-step is decided by one player, independently from the others.

The paper is organized as follows. In Section 2 we define some basic game-theoretical concepts extensively used along the paper. In Section 3 we introduce the game and analyze the quality of its equilibria. Then we explore the behavior on trees. In Section 4 we introduce a solution concept which allows coalitions of players. We show that this improves the Price of Anarchy, explore the computational complexity of finding equilibria and look at graphs with constant cut-width.

Finally, conclusions and directions for future work can found in Section 5. The omitted proofs can be found in an extended version of the paper [2].

2   Game-Theoretical Definitions

A strategic game $\mathcal{G} = (\mathcal{N}, \mathcal{S}_{i \in \mathcal{N}}, (u_i)_{i \in \mathcal{N}})$ is defined by a set of players $\mathcal{N}$, action sets $\mathcal{S}_i$ for each player $i \in \mathcal{N}$ and utilities $u_i : \mathcal{S} \to \mathbb{R}$, where $\mathcal{S} = \mathcal{S}_1 \times \ldots \times \mathcal{S}_{|\mathcal{N}|}$.

Each player i plays an action $s_i \in \mathcal{S}_i$ and his payoff is $u_i(s)$, where $s = (s_1, \ldots, s_{|\mathcal{N}|})$ is the strategy vector or strategy profile of all players. The quality of the outcome of the game when strategy vector s is played is measured by a so-called social welfare function $W(s)$. Furthermore we denote $(s_{-i}, s_i') = (s_1, \ldots, s_i', \ldots, s_{|\mathcal{N}|})$, i.e. strategy vector s, where player i changed his strategy from s_i to s_i'.

Nash Equilibrium. A strategy profile s is a Nash equilibrium, if no player can improve his payoff by changing the strategy he played. Let $\mathcal{E} \subseteq \mathcal{S}$ denote the set of all Nash equilibrium strategies. We say that $s \in \mathcal{E}$ if it holds that:

$$\forall i \in \mathcal{N}, \forall s_i' \in \mathcal{S}_i : u_i(s) \geq u_i(s_{-i}, s_i').$$

Price of Anarchy. The Price of Anarchy (PoA) of a game $\mathcal{G}$ with respect to a social welfare function W is defined as the ratio between the optimal solution and the worst equilibrium.

$$\mathrm{PoA}(\mathcal{G}, W) = \frac{\max_{s \in \mathcal{S}} W(s)}{\min_{s \in \mathcal{E}} W(s)}.$$

Price of Stability. The Price of Stability (PoS) of a game $\mathcal{G}$ with respect to a social welfare function W is defined as the ratio between the optimal solution and the best equilibrium.

$$\mathrm{PoS}(\mathcal{G}, W) = \frac{\max_{s \in \mathcal{S}} W(s)}{\max_{s \in \mathcal{E}} W(s)}.$$

3   The Firefighting Game

The Firefighting Problem takes place on an undirected graph $G = (V, E)$, where fire breaks out at one node, namely $v_0 \in V$, and incinerates all neighboring nodes at every time-step. We call those nodes *burning*. A fixed number b, called the budget, of firefighters can be placed on nodes to permanently protect them from burning. These nodes are called *defended*. If a node never burns because it is defended or cut off from the fire it is called *saved*. All other nodes are called *vulnerable*. We just consider the case of a $b = 1$.

In order to define a firefighting game, we have to define a set of players $\mathcal{N}$, with $\mathcal{N} = \{1, \ldots, n-1\}$ where $n = |V|$, and for every Player $i \in \mathcal{N}$, his strategy set $\mathcal{S}_i$ and his utility function u_i.

Player i decides which nodes to protect at time-step i. His strategy s_i is the subset of nodes he wants to place firefighters, $\mathcal{S}_i$ denotes the set of all possible strategies for player i. Since we only deal with the case of $b = 1$ we overload notation and instead of subsets of size one, we set the strategies to the vertices themselves or the empty set, i.e. $\mathcal{S}_i = V \cup \{\emptyset\}$. This means that players can choose one node or the empty set as a strategy. Let $s = (s_1, \ldots, s_{|\mathcal{N}|})$ denote the strategy profile of all players.

The outcome of the game is a partition of the vertex set into saved and burned nodes. It is defined in the following way. At time-step 0 the only burning node is v_0. At time-step $i > 0$, two events occur: First, player i's node is protected if his action is valid w.r.t. to strategy profile s, i.e. it is neither burning nor already defended at the end of time-step $i - 1$. Second, each node burning at time-step $i - 1$ incinerates all its non-defended neighbors. The process stops when the fire cannot spread any further. Let $\mathrm{Safe}(s) \subset V$ be the set of all nodes that are saved when strategy vector s is played. Furthermore, let $\mathrm{Safe}_i(s) = \mathrm{Safe}(s) \backslash \mathrm{Safe}(s_{-i}, \emptyset)$ be the set of nodes that would burn if player i switched his action to the empty set and let $\mathrm{invalid}(s, i)$ denote the event that player i's action is not valid with respect to strategy profile s.

3.1 Utility Functions

We look at two different functions, one modelling a selfish behavior and the other one modelling a non-profitable behavior. As it turns out, the respective games are equivalent.

a) Selfish Firefighters. In this model, firefighters get paid for the nodes they save. We call this game $\mathcal{G}^{(\mathrm{Selfish})}$. Intuitively, if player i makes a valid move other than the empty set, he gets one unit of currency from each node he helped to save. In other words, he gets paid by all nodes that are safe with respect to the played strategy vector, but would not be safe if he would change his strategy to the empty set. Additionally, he will get charged a penalty if he makes an invalid move. Now let us define the utility function formally.

$$u_i^{(\mathrm{Selfish})}(s) = \begin{cases} -c & \text{if } \mathrm{invalid}(s, i), \\ 0 & \text{if } s_i = \emptyset, \\ |\mathrm{Safe}_i(s)| - \varepsilon & \text{otherwise,} \end{cases}$$

with $0 < \varepsilon < 1$ and $c > 0$. We can see that the definition follows the intuition very closely. Subtracting an ε cost for placing a firefighter makes sure that players always prefer to play the empty set over placing a firefighter on a node that is already safe (which would not be an invalid move).

b) Non-Profit Firefighters. Here we assume that the goal of every firefighter is to save as many total nodes as possible, independently of which firefighters actually save more nodes. We call this game $\mathcal{G}^{\text{(Non-Profit)}}$. Formally, we define

$$u_i^{\text{(Non-Profit)}}(s) = \begin{cases} -c & \text{if invalid}(s, i), \\ |\text{Safe}(s)| & \text{if } s_i = \emptyset, \\ |\text{Safe}(s)| - \varepsilon & \text{otherwise,} \end{cases}$$

with $0 < \varepsilon < 1$ and $c > 0$.

Notice that in an equilibrium, no player plays an invalid move or puts a firefighter on an already safe node. Also, since we have that $0 < \varepsilon < 1$, the cost of placing a firefighter is less than the benefit of saving one node. Because of that, given that a player does not play the empty set, the ε-value does not affect his preferences. Therefore, we will ignore it in the proofs.

Equivalence of Games. Surprisingly, the behavior of selfish firefighters leads to the same equilibria than the behavior of the non-profit firefighters. It can be shown that the games $\mathcal{G}^{\text{(Selfish)}}$ and $\mathcal{G}^{\text{(Non-Profit)}}$ have the same sets of equilibria. This also implies that

$$\text{PoS}(\mathcal{G}^{\text{(Selfish)}}, W) = \text{PoS}(\mathcal{G}^{\text{(Non-Profit)}}, W)$$

$$\text{PoA}(\mathcal{G}^{\text{(Selfish)}}, W) = \text{PoA}(\mathcal{G}^{\text{(Non-Profit)}}, W).$$

Therefore we will use the utility function which is more convenient for the proof. Also, we will for now on refer to the game with $\mathcal{G}$, whenever the respective result holds for both versions of the game.

3.2   Quality of Equilibria

Once we have established a game, we can analyze the quality of the equilibria. In order to do this, we have to define a measure of the social benefit. We look at the simple case of the social welfare being the number of the nodes that are saved, i.e. $W(s) = |\text{Safe}(s)|$. It is easy to argue that equilibria always exist, because every optimal solution that does not contain invalid moves is an equilibrium for non-profit firefighter since it maximizes their utility function.

Price of Stability. In the case of non-profit firefighters, every strategy that maximizes the social welfare also maximizes the utility of every player given that he cannot improve his payoff by switching to the empty set. All optimal solution that are valid and do not protect nodes that are already saved are Nash equilibria. Therefore, we have the PoS is 1. This is independent of the class of graphs we are considering and holds for every solution concept where players maximize their utility function.

Lemma 1. $\text{PoS}(\mathcal{G}, W) = 1$. $\square$

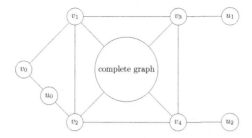

Fig. 1. Family of graphs $G_{PoA}(n) = (V_{PoA}(n), E_{PoA}(n))$. Note that $(v_1, v_4) \in E_{PoA}(n)$ and $(v_2, v_3) \in E_{PoA}(n)$. For better visibility these edges are not drawn in the picture. Further note that $|V_{PoA}(n)| = n$, hence the size of the complete graph is $n-8$ and the nodes of this graph together with nodes v_1, v_2, v_3 and v_4 form a clique of $G_{PoA}(n)$.

Price of Anarchy. In contrast to the PoS, the PoA is very high in this model. We first lower bound the PoA and then show that the bound is tight. For the proofs we use the utility functions of the selfish firefighters.

Theorem 1. $\text{PoA}(\mathcal{G}, W) \in \Theta(n)$.

Proof. We first prove a lower bound on the PoA, i.e. $\text{PoA}(\mathcal{G}, W) \in \Omega(n)$, and then show that this bound is tight. We look at an instance which has a very bad equilibrium relative to the optimal strategy with respect to the social welfare. Consider the family of graphs $G_{PoA}(n)$ shown in Figure 1.

Recall that the fire starts at v_0. It is easy to see that $s = (\{v_1\}, \{v_2\}, \emptyset^{n-3})$ is the optimal strategy. Only nodes v_0 and u_0 burn, hence the social welfare is $W(s) = n-2$. Furthermore we have that $s' = (\{v_3\}, \{v_4\}, \emptyset^{n-3})$ is an equilibrium. Note that the complete graph is burning after two time-steps, therefore at time-step 3 only u_1 and u_2 are neither burning nor defended. But these nodes are already safe, hence players i with $i > 2$ will not place firefighters on them. Furthermore, players 1 and 2 cannot improve their payoff, since if one of them changes strategy, that player will save at most one node. The social welfare of s' is $W(s') = 4$.

Hence, we have that $\text{PoA}(\mathcal{G}, W) \geq \frac{n-2}{4}$. It follows that $\text{PoA}(\mathcal{G}, W) \in \Omega(n)$. This means that we can only guarantee to save at most constant number of nodes. To argue that this bound is tight, we show that it is always possible to save a constant number of nodes.

By definition Player 1 can always place a firefighter on a node before the fire starts spreading. Also any strategy vector s where player 1 plays the empty set is not an equilibrium since he can always save at least one node which cannot be saved by any other player by placing a firefighter to a node adjacent to the original fire. This yields a upper bound of $\text{PoA}(\mathcal{G}, W) \leq n$, and hence $\text{PoA}(\mathcal{G}, W) \in \mathcal{O}(n)$. $\qquad\square$

3.3   Price of Anarchy for Trees

Since the PoA is very high in general, let us study the quality of equilibria for particular topologies. Our aim is to prove that there are cases where the quality of the equilibria is close to the quality of an optimal solution. In this section, we look at the PoA on trees. Let $\mathcal{G}_{\text{Tree}}$ denote the Firefighting Game on trees. We show that in contrast to our general result, the PoA is constant for trees. We assume that v_0, the initial fire, is the root of the tree.

Theorem 2. $PoA(\mathcal{G}_{Tree}, W) \leq 2$.

Proof. In this proof, we use similar ideas as in the proof of the approximation ratio of a greedy algorithm in a paper by Hartnell and Li [17].

We use the utility functions of the selfish firefighters. This implies that the utility of a player equals the size of the subtree he saves.

Let $\text{opt} = (\text{opt}_1, \dots, \text{opt}_{|\mathcal{N}|})$ be an optimal solution w.r.t to the social welfare, i.e. the optimal action opt_i is the node that is saved at time-step i. Let $s = (s_1, \dots, s_{|\mathcal{N}|})$ be an equilibrium strategy profile of the players. Recall that the optimal actions as well as the player actions are defined as the nodes in the tree that are saved. Let opt_A be the set of optimal actions opt_i, such that there is no player who plays the same action and no player action is an ancestor of opt_i, i.e. $\forall j \in \mathcal{N} : s_j \neq \text{opt}_i \wedge s_j$ is not ancestor of opt_i. Let opt_B denote the remaining optimal actions. Let $P(\text{opt}_i)$ denote the set of action s_j that are successors of opt_i. Let s_A denote the actions of players, that do not have an optimal action as an ancestor, i.e. $\forall j \in \mathcal{N} : \text{opt}_j$ is not ancestor of s_i. Let s_B denote the remaining player actions. Let $\text{save}(a)$ denote the numbers of nodes saved by action a.

Note that in opt_B there are optimal actions where a player plays the same action or a player action is an ancestor. Those corresponding player actions are the ones in s_A. Therefore we have that

$$\sum_{\text{opt}_i \in \text{opt}_B} \text{save}(\text{opt}_i) \leq \sum_{s_i \in s_A} \text{save}(s_i). \tag{1}$$

Because of the equilibrium property, we have that for every $\text{opt}_i \in \text{opt}_A$

$$\text{save}(s_i) \geq \text{save}(\text{opt}_i) - \sum_{s_j \in P(\text{opt}_i)} \text{save}(s_j),$$

because otherwise player i would have an incentive to switch his strategy to opt_i. If we now sum this up over all optimal actions in opt_A, we get

$$\sum_{\text{opt}_i \in \text{opt}_A} \text{save}(\text{opt}_i) \leq \sum_{\text{opt}_i \in \text{opt}_A} \left(\text{save}(s_i) + \sum_{s_j \in P(\text{opt}_i)} \text{save}(s_j) \right).$$

We can split up the sum on the left hand side and get $\sum_{\text{opt}_i \in \text{opt}_A} \text{save}(s_i) + \sum_{\text{opt}_i \in \text{opt}_A} \sum_{s_j \in P(\text{opt}_i)} \text{save}(s_j)$. Note that in the double sum, we sum up exactly over the player actions that have an optimal action as an ancestor i.e. s_B. So we can rewrite this to

$$\sum_{\text{opt}_i \in \text{opt}_A} \text{save}(\text{opt}_i) \leq \sum_{\text{opt}_i \in \text{opt}_A} \text{save}(s_i) + \sum_{s_i \in s_B} \text{save}(s_i).$$

Now we can use Inequality 1 to get

$$\sum_{\text{opt}_i \in \text{opt}} \text{save}(\text{opt}_i) \leq \sum_{\text{opt}_i \in \text{opt}_A} \text{save}(s_i) + \sum_{s_i \in s} \text{save}(s_i).$$

Furthermore, we have that $\sum_{\text{opt}_i \in \text{opt}_A} \text{save}(s_i) \leq \sum_{s_i \in s} \text{save}(s_i)$ which yields

$$\sum_{\text{opt}_i \in \text{opt}} \text{save}(\text{opt}_i) \leq 2 \sum_{s_i \in s} \text{save}(s_i).$$

This shows that an equilibrium strategy saves at least half of the nodes saved by an optimal solution, yielding a PoA of at most 2. □

4 Coalitions

In this section let us consider that players may form coalitions. A coalition is willing to deviate from their strategy as long as no player in the coalition loses payoff and at least one player increases his utility. We show that this affects the PoA. First, we need to introduce a suitable solution concept for coalitions.

We call a strategy vector s an equilibrium strategy with respect to coalition size k, if no set of at most k players can simultaneously change their strategies in such a way that at least one player increases his payoff and no player decreases his payoff. Let $K \subseteq \mathcal{N}$ denote the coalition and s_K a strategy profile of the members of the coalition. We say that coalition K has an *attractive joint deviation* if there is a strategy vector s'_K, such that $u_i(s) \leq u_i(s_{-K}, s'_K)$ for all $i \in K$, and for at least one player in K this inequality is strict.

Let $\mathcal{E}_k \subseteq \mathcal{S}$ denote the set of all equilibrium strategies with respect to coalition size k. We say that $s \in \mathcal{E}_k$, if there is no coalition K of size at most k that has an attractive joint deviation. Formally, we say that $s \in \mathcal{E}_k$ if it holds that:

$$\forall K \subseteq \mathcal{N} \text{ with } |K| \leq k \text{ and } \forall s'_K \neq s_K : s'_K \text{ is not an attractive joint deviation.}$$

Let $\mathcal{G}_k$ denote a firefighting game with coalitions of size at most k. In this case we do not have an equivalence between selfish and non-profit firefighters like in the Nash case. It can be shown that the sets of equilibria of the respective games are different. From now on we will only consider non-profit firefighters since they resemble the usual objective to save as many nodes as possible.

4.1 Price of Anarchy

Now we analyze the PoA for coalitions and its relation with the coalition size. We can show the following relationship.

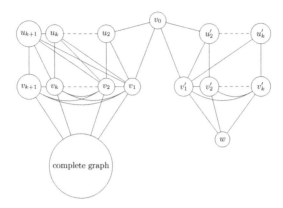

Fig. 2. Family of graphs $G_{PoA}(n,k) = (V_{PoA}(n,k), E_{PoA}(n,k))$, with $|V_{PoA}(n,k)| = n$. Note that the complete subgraph together with nodes v_1 to v_{k+1} form a clique. The nodes v'_1 to v'_k together with w form a clique as well. For every v_i and u_j and for every v'_i and u'_j there are edges (v_i, u_j) and (v'_i, u'_j), respectively, if $i \leq j$. Furthermore, for every u_i and u'_i there is an edge to u_{i+1} and u'_{i+1}, respectively.

Theorem 3. $PoA(\mathcal{G}_k, W) \in \Theta(\frac{n}{k})$.

Proof. To prove this, we first give an upper bound on the PoA for coalition size k. Later we show that this bound is tight. The upper bound we show is $PoA(\mathcal{G}_k, W) \leq \frac{n}{k} - 1$. We upper bound the welfare of the optimal solution and lower bound the welfare of the worst equilibrium. Note that if the optimal solution uses k or less time-steps, it can be found by a coalition of size k. Therefore, we assume that in the optimal solution at least in the first $k + 1$ time-steps a firefighter is placed on a node. This means that at most $n - k - 1$ nodes are saved. We can lower bound the number of nodes saved by the players by k, i.e. the nodes they place firefighters on. This yields a bound of the PoA of at most $\frac{n-k-1}{k} \leq \frac{n}{k} - 1$.

Now we show $PoA(\mathcal{G}_k, W) \geq \frac{n}{k+1} - 3$ for coalitions of size $k \leq \frac{n-3}{4}$.

We construct a family of graphs where the optimal solution saves at least all but $3k + 2$ nodes, whereas the worst equilibrium saves at most $k + 1$ nodes. Figure 2 shows the construction.

Note that any solution is a lower bound for the optimal solution and every equilibrium is an upper bound for the worst equilibrium in terms of quality.

The solution $s^* = (v_1, v_2, \ldots, v_{k+1}, \emptyset^{|\mathcal{N}|-k-1})$ saves all but $3k + 2$ nodes. This yields a lower bound for the welfare of an optimal solution.

Furthermore, we have that $s = (v'_1, v'_2, v'_3, \ldots, v'_k, \emptyset^{|\mathcal{N}|-k})$ is an equilibrium, since for every joint deviation the players can only save at most k nodes. In this equilibrium they save $k + 1$. Now we have a lower bound of the PoA of $\frac{n-3k-2}{k+1} \geq \frac{n}{k+1} - 3$.

Note that this construction uses at least $4k+3$ nodes, hence it is only applicable for coalition sizes up to $k \leq \frac{n-3}{4}$. Since the Price of Anarchy for size $k = \frac{n-3}{4}$ is constant this is no problem for the asymptotic bound.

We have bound the PoA from both sides and it follows that we have the claimed asymptotic behavior. □

It is interesting to see that for linear sized coalitions, we get a constant PoA. For constant coalition sizes however, the PoA is still linear. We can improve this result by fixing a special class of graphs, as we show in the next subsection.

4.2 Graphs with Constant Cut-Width

In this section we explore the impact of the cut-width of a graph on the Price of Anarchy for certain coalition sizes. We make use of results and ideas from Chlebíková and Chopin [7]. In particular, we show that for every family of graphs with constant cut-width there is a constant k, such that the PoA approaches one for coalitions of size k.

The cut-width of a graph G is defined as follows. The Cut-width $cw(G)$ of a graph G is the smallest integer k such that the vertices of G can be arranged in a linear layout $L = (v_0, \ldots, v_{n-1})$ in such a way that, for every $i \in \{0, \ldots, n-1\}$, there are at most k edges with one endpoint in $\{v_0, \ldots, v_i\}$ and the other in $\{v_{i+1}, \ldots, v_{n-1}\}$. Let $d_L(v_i, v_j) = |j - i|$ denote the distance between two nodes in the linear layout L.

Lemma 2. *If there is one initially burning node, then there exists a protection strategy such that the number of total burned nodes is at most $f(cw(G))$ for some function $f : \mathbb{N} \to \mathbb{N}$.* □

The proof of a more general version of this claim in contained in the proof of Theorem 2 of [7] and brings us into the position of showing the following lemma.

Lemma 3. *For every family of graphs $G(n) = (V(n), E(n))$ with constant cut-width there is a constant k, such that*

$$\lim_{n \to \infty} \text{PoA}(\mathcal{G}_k, W) = 1.$$

Proof. Let $G(n)$ be a family of graphs with constant cut-width. By Lemma 2 there is a protection strategy s, such that at most $f(cw(G))$ nodes burn. Now we make use of the fact that the number of time-steps before the spreading of the fire stops is less or equal to the total number of burned vertices. This is because in each time-step at least one node has to burn, otherwise the spreading of the fire would be stopped. Hence we get that with protection strategy s, the fire is contained in at most $f(cw(G))$ time-steps. Note that we can place at most one firefighter per time-step, therefore a coalition of size $k = f(cw(G))$ can apply this protection strategy. Furthermore, only a constant number of nodes burn. Hence, asymptotically, we have a PoA of 1. □

However, we cannot achieve this without coalitions as the following instance shows. Figure 3 shows a family of graphs. A linear layout is given by the horizontal position of the nodes in the figure. It shows that the cut-width of the

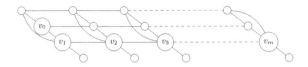

Fig. 3. Family of graphs with constant cut-width

graph is at most 6, since every vertical line through the graph crosses at most 6 edges. Without coalitions, saving the nodes v_1 to v_m is an equilibrium, since each player saves one extra node and cannot do better by switching to another node. Note that only a constant fraction of the nodes are saved, whereas in the case of coalition all nodes except a constant number can be saved. This also yields a constant PoA, but one that is asymptotically strictly larger than one.

This shows that for this class of graphs, constant sized coalitions can improve the PoA.

5   Conclusions

We have defined a new strategic game that models the Firefighter Problem. We have shown that in general PoA $\in \Theta(n)$. For trees however, we get a PoA of at most 2. Furthermore, we have shown that the coalition size has a direct effect on the quality of the equilibria. In general we have that PoA $\in \Theta(\frac{n}{k})$, where k is the coalition size. We have shown that there are topologies where PoA approaches 1 for constant sized coalitions, e.g. graphs with constant cut-width.

Note that it is possible to find equilibria in polynomial time for constant sized coalitions. This can be done by best response dynamics. Computing a best response is polynomial since we can try out all possible joint deviations for all possible coalitions of size at most k. With each best response the players impove the total number of saved nodes, hence we converge to an equilibrium in a linear number of iterations. This yields a polynomial time approximation algorithm for the firefighting problem and its approximation ratio equals the PoA of the corresponding game.

We think that the most promising area to explore is the quality of equilibria for other restricted sets of graphs. It is especially interesting to find sets of graphs that have a low PoA for constant sized coalitions.

References

1. Alon, N., Feldman, M., Procaccia, A.D., Tennenholtz, M.: A note on competitive diffusion through social networks. Information Processing Letters **110**, 221–225 (2010)
2. Àlvarez, C., Blesa, M., Molter, H.: Firefighting as a Game. Technical Report LSI-14-9-R, Computer Science Dept, Universitat Politècnica de Catalunya (2014)
3. Anshelevich, E., Chakrabarty, D., Hate, A., Swamy, C.: Approximability of the firefighter problem. Algorithmica **62**, 520–536 (2012)

4. Bazgan, C., Chopin, M., Cygan, M., Fellows, M.R., Fomin, F., Jan van Leeuwen, E.: Parameterized complexity of firefighting. Journal of Computer and System Sciences **80**, 1285–1297 (2014)
5. Bazgan, C., Chopin, M., Ries, B.: The firefighter problem with more than one firefighter on trees. Discrete Applied Mathematics **161**, 899–908 (2013)
6. Cai, L., Verbin, E., Yang, L.: Firefighting on Trees: $(1 - 1/e)$–Approximation, Fixed Parameter Tractability and a Subexponential Algorithm. In: Hong, S.-H., Nagamochi, H., Fukunaga, T. (eds.) ISAAC 2008. LNCS, vol. 5369, pp. 258–269. Springer, Heidelberg (2008)
7. Chlebíková, J., Chopin, M.: The firefighter problem: A structural analysis. Electronic Colloquium on Computational Complexity **20**, 162 (2013)
8. Costa, V., Dantas, S., Dourado, M.C., Penso, L., Rautenbach, D.: More fires and more fighters. Discrete Applied Mathematics **161**, 2410–2419 (2013)
9. Cygan, M., Fomin, F.V., van Leeuwen, E.J.: Parameterized Complexity of Firefighting Revisited. In: Marx, D., Rossmanith, P. (eds.) IPEC 2011. LNCS, vol. 7112, pp. 13–26. Springer, Heidelberg (2012)
10. Feldheim, O.N., Hod, R.: 3/2 Firefighters Are Not Enough. Discrete Applied Mathematics **161**, 301–306 (2013)
11. Finbow, S., King, A., MacGillivray, G., Rizzi, R.: The firefighter problem for graphs of maximum degree three. Discrete Mathematics **307**, 2094–2105 (2007)
12. Finbow, S., MacGillivray, G.: The Firefighter Problem: A survey of results, directions and questions. Australian Journal of Combinatorics **43**, 57–77 (2009)
13. Floderus, P., Lingas, A., Persson, M.: Towards more efficient infection and fire fighting. International Journal of Foundations of Computer Science **24**, 3–14 (2013)
14. Fomin, F.V., Heggernes, P., van Leeuwen, E.J.: Making Life Easier for Firefighters. In: Kranakis, E., Krizanc, D., Luccio, F. (eds.) FUN 2012. LNCS, vol. 7288, pp. 177–188. Springer, Heidelberg (2012)
15. Grötschel, M., Lovász, L., Schrijver, A.: Geometric Algorithms and Combinatorial Optimization. Springer (1988)
16. Hartnell, B.: Firefighter! an application of domination. In: 25th Manitoba Conference on Combinatorial Mathematics and Computing, University of Manitoba in Winnipeg, Canada (1995)
17. Hartnell, B., Li, Q.: Firefighting on trees: How bad is the greedy algorithm? Congressus Numerantium **145**, 187–192 (2000)
18. Iwaikawa, Y., Kamiyama, N., Matsui, T.: Improved Approximation Algorithms for Firefighter Problem on Trees. IEICE Transactions **94-D**, 196–199 (2011)
19. King, A., MacGillivray, G.: The firefighter problem for cubic graphs. Discrete Mathematics **310**, 614–621 (2010)
20. MacGillivray, G., Wang, P.: On the firefighter problem. Journal of Combinatorial Mathematics and Combinatorial Computing **47**, 83–96 (2003)
21. Ng, K., Raff, P.: A generalization of the firefighter problem on Z × Z. Discrete Applied Mathematics **156**, 730–745 (2008)
22. Small, L., Mason, O.: Nash Equilibria for competitive information diffusion on trees. Information Processing Letters **113**, 217–219 (2013)
23. Small, L., Mason, O.: Information diffusion on the iterated local transitivity model of online social networks. Discrete Applied Mathematics **161**, 1338–1344 (2013)
24. Takehara, R., Hachimori, M., Shigeno, M.: A comment on pure-strategy Nash equilibria in competitive diffusion games. Information Processing Letters **112**, 59–60 (2012)
25. Zinoviev, D., Duong, V., Zhang, H.: A Game Theoretical Approach to Modeling Information Dissemination in Social Networks. CoRR, abs/1006.5493 (2010)

PageRank in Scale-Free Random Graphs

Ningyuan Chen[1], Nelly Litvak[2], and Mariana Olvera-Cravioto[1]([✉])

[1] Columbia University, 500 W. 120th Street, 3rd floor, New York, NY 10027, USA
mo2291@columbia.edu
[2] University of Twente, P.O.Box 217, 7500 AE Enschede, The Netherlands

Abstract. We analyze the distribution of PageRank on a directed configuration model and show that as the size of the graph grows to infinity, the PageRank of a randomly chosen node can be closely approximated by the PageRank of the root node of an appropriately constructed tree. This tree approximation is in turn related to the solution of a linear stochastic fixed-point equation that has been thoroughly studied in the recent literature.

1 Introduction

Google's PageRank proposed by Brin and Page [4] is arguably the most influential technique for computing centrality scores of nodes in a network, see [10] for a thorough review. In this paper we analyze the power law behavior of PageRank scores in scale-free directed random graphs.

In real-world networks, it is often found that the fraction of nodes with (in- or out-) degree k is $\approx c_0 k^{-\alpha-1}$, usually $\alpha \in (1,3)$, see e.g. [14] for an excellent review of the mathematical properties of complex networks.

More than ten years ago Pandurangan et al. [13] discovered the interesting fact that PageRank scores also exhibit power laws, with the same exponent as the in-degree. This property holds for a broad class of real-life networks [16]. In fact, the hypothesis that this always holds in power-law networks is plausible. However, analytical mathematical evidence supporting this hypothesis is surprisingly scarce. As one of the few examples, Avrachenkov and Lebedev [3] obtained the power law behavior of average PageRank scores in a preferential attachment graph by using Polya's urn scheme and advanced symbolic computations.

In a series of papers, Volkovich et al. [11,15,16] suggested an analytical explanation for the power law behavior of PageRank by comparing it to the endogenous solution of a stochastic fixed-point equation (SFPE). The properties of this equation and the study of its multiple solutions has itself been an interesting topic in the recent literature [1,2,7–9,12], and is related to the broader study of weighted branching processes. The tail behavior of the endogenous solution, the one more closely related to PageRank, was given in [7–9,12], where it was shown to have a power law under many different sets of assumptions. However,

The second author was partially funded by the EU-FET Open grant NADINE (288956). The third author was supported by the NSF grant CMMI-1131053.

© Springer International Publishing Switzerland 2014
A. Bonato et al. (Eds.): WAW 2014, LNCS 8882, pp. 120–131, 2014.
DOI: 10.1007/978-3-319-13123-8_10

the SFPE does not fully explain the behavior of PageRank in networks since it implicitly assumes that the underlying graph is an infinite tree, an assumption that is not in general satisfied in real-world networks.

This paper makes a fundamental step further by extending the analysis of PageRank to graphs that are not necessarily trees. Specifically, we analyze PageRank in a directed configuration model (DCM) with given degree distributions, as developed by Chen and Olvera-Cravioto [6]. We present numerical evidence that in this type of graphs the behavior of PageRank is very close to the one on trees. Intuitively, this is true for two main reasons: 1) the influence of remote nodes on the PageRank of an arbitrary node decreases exponentially fast with the graph distance; and 2) the DCM is asymptotically tree-like, that is, when we explore a graph starting from a given node, then with high probability the first loop is observed at a distance of order $\log n$, where n is the size of the graph (see Figure 1).

Our main result establishes analytically that PageRank in a DCM is well approximated by the PageRank of the root node of a suitably constructed tree as the graph size goes to infinity. As a consequence, the analysis of PageRank on the graph reduces to studying PageRank on a tree, a problem that, as mentioned earlier, can be solved by using the properties of the SFPE. In particular, since the endogenous solution to the SFPE is known to have a power-law tail when the in-degree follows a power-law, our main result allows us to establish the power-law behavior of PageRank on the graph.

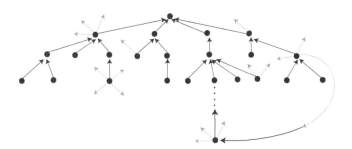

Fig. 1. Graph construction process. Unpaired outbound stubs are in blue.

Section 2 below describes the DCM as presented in [6]. Then, in Section 3 we analytically compare the PageRank scores in the DCM to their approximate value obtained after a finite number of power iterations. Next, in Section 4 we explain how to couple the PageRank of a randomly chosen node with the root node of a suitable branching tree, and give our main analytical results. Finally, in Section 5 we give numerical results validating our analytical work. The complete proofs for more general stochastic recursions, that also cover the PageRank case considered here, are given in [5], which also contains a detailed presentation of the corresponding SFPEs and the results that can be derived from there.

2 Directed Random Graphs

We will give below an algorithm, taken from [6], that can be used to generate a scale-free directed graph. Formally, power law distributions are modeled using the mathematical notion of regular variation. A nonnegative random variable X is said to be regularly varying, if $\overline{F}(x) := P(X > x) = L(x)x^{-\alpha}$, $x > 0$, where $L(\cdot)$ is a slowly varying function, that is, $\lim_{x \to \infty} L(tx)/L(x) = 1$, for all $t > 0$.

Our goal now is to create a directed graph $\mathcal{G}_n$ with the property that the in-degrees and out-degrees will be approximately distributed, for large sizes of the graph, according to distributions $f_k^{in} = P(\mathcal{N} = k)$, and $f_k^{out} = P(\mathcal{D} = k)$, $k = 0, 1, 2, 3, \ldots$, respectively, where $E[\mathcal{N}] = E[\mathcal{D}]$. The only condition needed is that these distributions satisfy

$$\overline{F^{in}}(x) = \sum_{k>x} f_k^{in} \leq x^{-\alpha} L_{in}(x) \quad \text{and} \quad \overline{F^{out}}(x) = \sum_{k>x} f_k^{out} \leq x^{-\beta} L_{out}(x),$$

for some slowly varying functions $L_{in}(\cdot)$ and $L_{out}(\cdot)$, and $\alpha, \beta > 1$.

The first step in our procedure is to generate an appropriate bi-degree sequence

$$(\mathbf{N}_n, \mathbf{D}_n) = \{(N_i, D_i) : 1 \leq i \leq n\}$$

representing the n nodes in the graph. The algorithm given below will ensure that the in- and out-degrees follow closely the desired distributions and also that the sums of in- and out-degrees are the same:

$$L_n := \sum_{i=1}^{n} N_i = \sum_{i=1}^{n} D_i.$$

Denote

$$\kappa_0 = \min\{1 - \alpha^{-1}, 1 - \beta^{-1}, 1/2\}.$$

Algorithm 1. *Generation of a bi-degree sequence with given in-/out-degree distributions.*

1. Fix $0 < \delta_0 < \kappa_0$.
2. Sample an i.i.d. sequence $\{\mathcal{N}_1, \ldots, \mathcal{N}_n\}$ from distribution F^{in}.
3. Sample an i.i.d. sequence $\{\mathcal{D}_1, \ldots, \mathcal{D}_n\}$ from distribution F^{out}, independent of $\{\mathcal{N}_i\}$.
4. Define $\Delta_n = \sum_{i=1}^{n} (\mathcal{N}_n - \mathcal{D}_n)$. If $|\Delta_n| \leq n^{1-\kappa_0+\delta_0}$ proceed to step 5; otherwise repeat from step 2.
5. Choose randomly $|\Delta_n|$ nodes $\{i_1, i_2, \ldots, i_{|\Delta_n|}\}$ without replacement and let

$$N_i = \begin{cases} \mathcal{N}_i + 1 & \text{if } \Delta_n < 0 \text{ and } i \in \{i_1, i_2, \ldots, i_{|\Delta_n|}\}, \\ \mathcal{N}_i & \text{otherwise}, \end{cases}$$

$$D_i = \begin{cases} \mathcal{D}_i + 1 & \text{if } \Delta_n \geq 0 \text{ and } i \in \{i_1, i_2, \ldots, i_{|\Delta_n|}\}, \\ \mathcal{D}_i & \text{otherwise}. \end{cases}$$

Remark: It was shown in [6] that

$$P\left(|\Delta_n| > n^{1-\kappa_0+\delta_0}\right) = O\left(n^{-\delta_0(\kappa_0-\delta_0)/(1-\kappa_0)}\right) \tag{1}$$

as $n \to \infty$, and therefore Algorithm 1 will always terminate after a finite number of steps (i.e., it will eventually proceed to step 5).

Having obtained a realization of the bi-degree sequence $(\mathbf{N}_n, \mathbf{D}_n)$, we now use the configuration model to construct the random graph. The idea in the directed case is essentially the same as for undirected graphs. To each node v_i we assign N_i inbound half-edges and D_i outbound half-edges; then, proceed to match inbound half-edges to outbound half-edges to form directed edges. To be more precise, for each unpaired inbound half-edge of node v_i choose randomly from all the available unpaired outbound half-edges, and if the selected outbound half-edge belongs to node, say, v_j, then add a directed edge from v_j to v_i to the graph; proceed in this way until all unpaired inbound half-edges are matched. Note that the resulting graph is not necessarily simple, i.e., it may contain self-loops and multiple edges in the same direction.

We point out that conditional on the graph being simple, it is uniformly chosen among all simple directed graphs having bi-degree sequence $(\mathbf{N}_n, \mathbf{D}_n)$ (see [6]). Moreover, it was also shown in [6] that, provided $\alpha, \beta > 2$, the probability of obtaining a simple graph through this procedure is bounded away from zero, and therefore one can obtain a simple graph having $(\mathbf{N}_n, \mathbf{D}_n)$ as its bi-degree sequence by simply repeating the algorithm enough times. When we can only ensure that $\alpha, \beta > 1$, then a simple graph can still be obtained without loosing the distributional properties of the in- and out-degrees by erasing the self-loops and merging multiple edges in the same direction. These considerations about the graph being simple are nonetheless irrelevant to the ranking problem here.

3   PageRank Iterations in the DCM

Although PageRank can be thought of as the solution to a system of linear equations, we will show in this section how it is sufficient to consider only a finite number of matrix iterations to obtain an accurate approximation for the PageRank of all the nodes in the graph. We first introduce some notation.

Let $M = M(n) \in \mathbb{R}^{n \times n}$ be the matrix constructed as follows:

$$M_{i,j} = \begin{cases} s_{ij}c/D_i, & \text{if there are } s_{ij} \text{ edges from } i \text{ to } j, \\ 0, & \text{otherwise,} \end{cases}$$

and let $\mathbf{1}$ be the row vector of ones. In the classical definition [10], PageRank $\pi = (\pi_1, \ldots, \pi_n)$ is the unique solution to the following equation:

$$\pi = \pi(cM) + \frac{1-c}{n}\mathbf{1}, \tag{2}$$

where $c \in (0,1)$ is a parameter known as the damping factor. Rather than analyzing π directly, we consider instead its scale-free version

$$n\pi =: \mathbf{R} = \mathbf{R}(cM) + (1-c)\mathbf{1} \tag{3}$$

obtained by multiplying (2) by the size of the graph n. Moreover, whereas π_i is a probability distribution ($\pi_i \geq 0$ for all i and $\pi\mathbf{1}^T = 1$), its scale-free version $\mathbf{R} = (R_1, \ldots, R_n)$ has components that are essentially unbounded for large n and that satisfy $E[R_i] = 1$ for all $1 \leq i \leq n$ and all n (hence the name *scale-free*).

One way to solve the system of linear equations given in (3) is via power iterations. We define the kth iteration of PageRank on the graph as follows. First initialize PageRank with a vector $\mathbf{r}_0 = r_0\mathbf{1}$, $r_0 \geq 0$, and then iterate according to $\mathbf{R}^{(n,0)} = \mathbf{r}_0$ and

$$\mathbf{R}^{(n,k)} = \mathbf{R}^{(n,k-1)}M + (1-c)\mathbf{1} = (1-c)\mathbf{1}\sum_{i=0}^{k-1} M^i + \mathbf{r}_0 M^k$$

for $k \geq 1$. In this notation, $\mathbf{R} = \mathbf{R}^{(n,\infty)}$, and our main interest is to analyze the distribution of the PageRank of a randomly chosen node in the DCM, say $R_1^{(n,\infty)}$. The first step of the analysis is to compare $\mathbf{R}^{(n,\infty)}$ to its kth iteration $\mathbf{R}^{(n,k)}$. To this end, note that $\mathbf{R}^{(n,\infty)} = (1-c)\mathbf{1}\sum_{i=0}^{\infty} M^i$, and therefore,

$$\mathbf{R}^{(n,k)} - \mathbf{R}^{(n,\infty)} = \mathbf{r}_0 M^k - (1-c)\mathbf{1}\sum_{i=k}^{\infty} M^i.$$

Moreover,

$$\left\|\mathbf{R}^{(n,k)} - \mathbf{R}^{(n,\infty)}\right\|_1 \leq \left\|\mathbf{r}_0 M^k\right\|_1 + (1-c)\sum_{i=0}^{\infty} \left\|\mathbf{1}M^{k+i}\right\|_1$$

$$\leq r_0 n \left\|M^k\right\|_\infty + (1-c)n\sum_{i=0}^{\infty} \left\|M^{k+i}\right\|_\infty,$$

where for the last inequality we used the observation that

$$\left\|\mathbf{1}M^r\right\|_1 = \sum_{j=1}^{n}\sum_{i=1}^{n}(M^r)_{ij} = \sum_{i=1}^{n} \left\|(M^r)_{i\bullet}\right\|_1 \leq n\left\|M^r\right\|_\infty,$$

where $A_{i\bullet}$ denotes the ith row of matrix A. Furthermore, since M is equal to c times a transition probability matrix, we have

$$\left\|M^r\right\|_\infty \leq \left\|M\right\|_\infty^r = c^r.$$

It follows that

$$\left\|\mathbf{R}^{(n,k)} - \mathbf{R}^{(n,\infty)}\right\|_1 \leq r_0 n c^k + (1-c)n\sum_{i=0}^{\infty} c^{k+i} = (r_0+1)nc^k. \tag{4}$$

The approach used to derive bound (4) for the L_1-norm of the error is valid for any directed network. However, this bound does not, in the general case, provide information on the convergence of specific coordinates and does not give a good upper bound for the quantity $|R_1^{(n,k)} - R_1^{(n,\infty)}|$ that we are interested in. It is here where the structure of the DCM plays a role, since by construction, it makes all permutations of the nodes' labels equally likely, which implies that all coordinates of the vector $\mathbf{R}^{(n,k)} - \mathbf{R}^{(n,\infty)}$ have the same distribution. This leads to the following observation.

Let $\mathscr{F}_n = \sigma((\mathbf{N}_n, \mathbf{D}_n))$ denote the sigma-algebra generated by the bi-degree sequence, which does not include information about the pairing process. Then, conditional on $\mathscr{F}_n$,

$$E\left[\left|R_1^{(n,k)} - R_1^{(n,\infty)}\right| \,\Big|\, \mathscr{F}_n\right] = \frac{1}{n} E\left[\left\|\mathbf{R}^{(n,k)} - \mathbf{R}^{(n,\infty)}\right\|_1 \,\Big|\, \mathscr{F}_n\right] \leq (r_0 + 1)\, c^k,$$

and for any $\epsilon > 0$ Markov's inequality gives,

$$P\left(\left|R_1^{(n,\infty)} - R_1^{(n,k)}\right| > \epsilon\right) \leq E\left[\epsilon^{-1} E\left[\left|R_1^{(n,k)} - R_1^{(n,\infty)}\right| \,\Big|\, \mathscr{F}_n\right]\right]$$
$$\leq (r_0 + 1)\, \epsilon^{-1} c^k. \tag{5}$$

Note that (5) is a probabilistic statement, which is not completely analogous to (4). In fact, (5) states that we can achieve any level of precision with a pre-specified high probability by simply increasing the number of iterations k. This leads to the following heuristic, that if the DCM looks locally like a tree for k generations, where k is the number of iterations needed to achieve the desired precision in (5), then the PageRank of node 1 in the DCM will be essentially the same as the PageRank of the root node of a suitably constructed tree. The precise result and a sketch of the arguments will be given in the next section.

4   Main Result: Coupling with a Thorny Branching Tree

As mentioned in the previous section, we will now show how to identify $R_1^{(n,k)}$ with the PageRank of the root node of a tree. To start, we construct a variation of a branching tree where each node has an edge pointing to its parent but also has a number of outbound stubs or half-edges that are pointing outside of the tree (i.e., to some auxiliary node). We will refer to this tree as a Thorny Branching Tree (TBT), the name "thorny" referring to the outbound stubs (see Figure 1).

To construct simultaneously the graph $\mathcal{G}_n$ and the TBT, denoted by $\mathcal{T}$, we start by choosing a node uniformly at random, and call it node 1 (the root node). This first node will have N_1 inbound stubs which we will proceed to match with randomly chosen outbound stubs. These outbound stubs are sampled independently and with replacement from all the possible $L_n = \sum_{i=1}^{n} D_i$ outbound stubs, discarding any outbound stub that has already been matched.

This corresponds to drawing independently at random from the distribution

$$f_n(i,j) = P(\text{node has } i \text{ offspring, } j \text{ outbound links } |\mathscr{F}_n)$$

$$= \sum_{k=1}^{n} 1(N_k = i, D_k = j) P(\text{an outbound stub of node } k \text{ is sampled } |\mathscr{F}_n)$$

$$= \sum_{k=1}^{n} 1(N_k = i, D_k = j) \frac{D_k}{L_n}. \tag{6}$$

This is a so-called size-biased distribution, since nodes with more outbound stubs are more likely to be chosen.

To keep track of which outbound stubs have already been matched we will label them 1, 2, or 3 according to the following rule:

1. Outbound stubs with label 1 are stubs belonging to a node that is not yet attached to the graph.
2. Outbound stubs with label 2 belong to nodes that are already part of the graph but that have not yet been paired with an inbound stub.
3. Outbound stubs with label 3 are those which have already been paired with an inbound stub and now form an edge in the graph.

Let Z_r, $r \geq 0$, denote the number of inbound stubs of all the nodes in the graph at distance r of the first node. Note that $Z_0 = N_1$ and Z_r is also the number of nodes at distance $(r+1)$ of the first node.

To draw the graph we initialize the process by labeling all outbound stubs with a 1, except for the D_1 outbound stubs of node 1 that receive a 2. We then start by pairing the first of the N_1 inbound stubs with a randomly chosen outbound stub, say belonging to node j. Then node j is attached to the graph by forming an edge with node 1, and all the outbound stubs from the new node are now labeled 2. In case that $j = 1$ the pairing forms a self-loop and no new nodes are added to the graph. Next, we label the chosen outbound stub with a 3, since it has already been paired, and in case $j \neq 1$, give all the other outbound stubs of node j a label 2. We continue in this way until all N_1 inbound stubs of node 1 have been paired, after which we will be left with Z_1 unmatched inbound stubs that will determine the nodes at distance 2 from node 1. In general, the kth iteration of this process is completed when all Z_{k-1} inbound stubs have been matched with an outbound stub, and the process ends when all L_n inbound stubs have been paired. Note that whenever an outbound stub with label 2 is chosen a cycle or double edge is formed in the graph. If at any point we sample an outbound stub with label 3 we simply discard it and do a redraw until we obtain an outbound stub with labels 1 or 2.

We now explain the coupling with the TBT. We start with the root node (node 1, generation 0) that has $\hat{N}_1 = N_1$ offspring. Let $\hat{Z}_k$ denote the number of individuals in generation $k + 1$ of the tree, $\hat{Z}_0 = \hat{N}_1$. For $k \geq 1$, each of the $\hat{Z}_{k-1}$ individuals in the kth generation will independently have offspring and outbound stubs according to the random joint distribution $f_n(i,j)$ given in (6).

The coupling of $\mathcal{G}_n$ and the TBT is done according to the following rules:

1. If an outbound stub with label 1 is chosen, then both the graph and the TBT will connect the chosen outbound stub to the inbound stub being matched, resulting in a node being added to the graph and an offspring being born to its parent. In particular, if the chosen outbound stub corresponds to node j, then the new offspring in the TBT will have $D_j - 1$ outbound stubs (pointing to the auxiliary node) and N_j inbound stubs (number of offspring). We then update the labels by giving a 2 label to all the 'sibling' outbound stubs of the chosen outbound stub, and a 3 label to the chosen outbound stub itself.
2. If an outbound stub with label 2 is sampled it means that its corresponding node already belongs to the graph, and a cycle, self-loop, or multiple edge is created. In $\mathcal{T}$, we proceed as if the outbound stub had label 1 and create a new node, which is a copy of the drawn node. The coupling between DCM and TBT breaks at this point.
3. If an outbound stub with label 3 is drawn it means that this stub has already been matched, and the coupling breaks as well. In $\mathcal{T}$, we again proceed as if the outbound stub had had a label 1. In the graph we do a redraw.

Note that the processes Z_k and $\hat{Z}_k$ are identical as long as the coupling holds. Showing that the coupling holds for a sufficient number of generations is the essence of our main result.

Definition 1. *Let τ be the number of generations in the TBT that can be completed before the first outbound stub with label 2 or 3 is drawn, i.e., $\tau = k$ iff the first inbound stub to draw an outbound stub with label 2 or 3 belonged to a node i, such that the graph distance between i and the root node is exactly k.*

The following result gives us an estimate as to when the coupling between the exploration process of the graph and the construction of the tree is expected to break.

Lemma 1. *Suppose $(\mathbf{N}_n, \mathbf{D}_n)$ are constructed using Algorithm 1 with $\alpha > 1$, and $\beta > 2$. Let $\mu = E[\mathcal{N}] = E[\mathcal{D}] > 1$. Then, for any $1 \leq k \leq h \log n$ with $0 < h < 1/(2 \log \mu)$ there exists a $\delta > 0$ such that,*

$$P(\tau \leq k) = O\left(n^{-\delta}\right) \qquad \text{as } n \to \infty.$$

The proof of Lemma 1 is rather technical, so we will only provide a sketch in this paper. The detailed proof is given in [5].

Proof (Qualitative argument). Let $\hat{V}_s$ be the number of outbound stubs of all nodes in generation s of the tree. The intuition behind the proof is that for all $s = 1, 2, \ldots$, neither $\hat{Z}_s$, nor $\hat{V}_s$ are expected to be much larger than their means:

$$E\left[\hat{Z}_s \middle| \mathscr{F}_n\right] \approx \mu^{s+1} \qquad \text{and} \qquad E\left[\hat{V}_s \middle| \mathscr{F}_n\right] \approx \lambda \mu^s,$$

where $\lambda = E[\mathscr{D}^2]/\mu$. Next, note that an inbound stub of a node in the rth generation will be the first one to be paired with an outbound stub having label 2 or 3 with a probability bounded from above by

$$P_r := \frac{1}{L_n} \sum_{s=0}^{r} \hat{V}_s \approx \frac{\lambda \mu^r}{n(\mu - 1)}.$$

Furthermore, for event $\{\tau = r\}$ to occur one of the $\hat{Z}_r$ inbound stubs must have been paired with an outbound stub with labels 2 or 3, which is bounded by the probability that a Binomial random variable with parameters $(\hat{Z}_r, P_r)$ is greater or equal than 1. By Markov's inequality we then have that this probability is smaller or equal than $\hat{Z}_r P_r = O\left(\mu^{2r} n^{-1}\right)$ for $r \leq k$.

Formally, to ensure that the approximations given above are valid, we first show that the event

$$E_k = \left\{ \max_{0 \leq r \leq k} \mu^{-r} \hat{Z}_r \leq x_n \right\}$$

occurs with high probability as $n \to \infty$ for a suitably chosen $x_n \to \infty$. Then, sum over $r = 0, 1, \ldots, k$ the events $\{\tau = r, E_k\}$ to obtain that $P(\tau \leq k, E_k) = O\left(\mu^{2k} n^{-1}\right)$, which goes to zero for $k \leq h \log n$.

Our main result is now a direct consequence of the bound derived in (5) and Lemma 1 above, since before the coupling breaks $R_1^{(n,k)}$ and the PageRank, computed after k iterations, of the root node of the coupled tree coincide.

Theorem 1. *Suppose* $(\mathbf{N}_n, \mathbf{D}_n)$ *are constructed using Algorithm 1 with* $\alpha > 1$, *and* $\beta > 2$. *Let* $\mu = E[\mathscr{N}] = E[\mathscr{D}] > 1$ *and* $c \in (0, 1)$. *Then, for any* $\epsilon > 0$ *and any* $1 \leq k \leq h \log n$ *with* $0 < h < 1/(2 \log \mu)$ *there exists a* $\delta > 0$ *such that,*

$$P\left(\left| R_1^{(n,\infty)} - \hat{R}_1^{(n,k)} \right| > \epsilon \right) \leq (r_0 + 1)\epsilon^{-1} c^k + O\left(n^{-\delta}\right),$$

as $n \to \infty$, *where* $\hat{R}_1^{(n,k)}$ *is the PageRank, after* k *iterations, of the root node of the TBT described above.*

In [5] we explore further the distribution of the PageRank of the root node of $\mathcal{T}$ and show that $\hat{R}_1^{(n,k)}$ converges to the endogenous solution of a SFPE on a weighted branching tree, as originally suggested in [11,15,16]. Moreover, the tail behavior of this solution has been fully described in [7,8,15].

5 Numerical Results

In this last section we give some numerical results showing the accuracy of the TBT approximation to the PageRank in the DCM. To generate the bi-degree sequence we use as target distributions two Pareto-like distributions. More precisely, we set

$$\mathscr{N}_i = \lfloor X_{1,i} + Y_{1,i} \rfloor, \quad \mathscr{D}_i = \lfloor X_{2,i} + Y_{2,i} \rfloor,$$

where the $\{X_{1,i}\}$ and the $\{X_{2,i}\}$ are independent sequences of i.i.d. Pareto random variables with shape parameters $\alpha > 1$ and $\beta > 2$, respectively, and scale parameters $x_1 = (\alpha - 1)/\alpha$ and $x_2 = (\beta - 1)/\beta$, respectively (note that $E[X_{1,i}] = E[X_{2,i}] = 1$ for all i). The sequences $\{Y_{1,i}\}$ and $\{Y_{2,i}\}$ are independent sequences, each consisting of i.i.d. exponential random variables with means $1/\lambda_1 > 0$ and $1/\lambda_2$, respectively. The addition of the exponential random variables allows more flexibility in the modeling of the in- and out-degree distributions while preserving a power law tail behavior; the parameters λ_1, λ_2 are also used to match the means $E[\mathcal{N}]$ and $E[\mathcal{D}]$.

Once the sequences $\{\mathcal{N}_i\}$ and $\{\mathcal{D}_i\}$ are generated, we use Algorithm 1 to obtain a valid bi-degree sequence $(\mathbf{N}_n, \mathbf{D}_n)$. Given this bi-degree sequence we next proceed to construct the graph and the TBT simultaneously, according to the rules described in Section 4. To compute $\mathbf{R}^{(n,\infty)}$ we perform matrix iterations with $r_0 = 1$ until $\|\mathbf{R}^{(n,k)} - \mathbf{R}^{(n,k-1)}\|_2 < \epsilon_0$ for some tolerance ϵ_0. We only generate the TBT for the required number of generations in each of the examples; the computation of $\hat{R}_1^{(n,k)}$ can be done recursively starting from the leaves using

$$\hat{R}_i^{(n,0)} = 1, \quad \hat{R}_i^{(n,k)} = \sum_{j \to i} c\hat{R}_j^{(n,k-1)} + (1 - c), \quad k > 0, \tag{7}$$

where $j \to i$ means that node j is an offspring of node i. We use $\|\cdot\|_2$ here in order to provide mean squared errors (MSEs) for our approximations.

Tables 1-3 below compare the PageRank of node 1 in the graph, $R_1^{(n,\infty)}$, the PageRank of node 1 only after k power iterations, $R_1^{(n,k)}$, and the PageRank of the root node of the coupled tree after the same k generations, $\hat{R}_1^{(n,k)}$. The magnitude of the MSEs, computed using $R_1^{(n,\infty)}$ as the true value, is also given in each table. The tolerance for computing $R_1^{(n,\infty)}$ is set to $\varepsilon_0 = 10^{-6}$. For each n, we generate 100 realizations of $\mathcal{G}_n$ as well as of the corresponding TBTs and take the empirical average of the PageRank values and of the MSEs. Table 1 includes results for different sizes of the graph, and uses $k_n = \lfloor \log n \rfloor$ iterations for the finite approximations. We note that all the MSEs clearly decrease as n increases since k_n also increases with n.

Table 1. $\alpha = 2, \beta = 2.5, \lambda_1 = 1, c = 0.5, k_n = \lfloor \log n \rfloor$

n	$R_1^{(n,\infty)}$	$R_1^{(n,k_n)}$	$\hat{R}_1^{(n,k_n)}$	MSE for $R_1^{(n,k_n)}$	MSE for $\hat{R}_1^{(n,k_n)}$
10	0.931	0.946	0.983	3.90E-03	4.20E-02
100	1.023	1.027	1.068	1.80E-04	3.70E-02
1000	1.000	1.002	1.010	1.20E-05	8.00E-04
10000	0.964	0.965	0.962	1.00E-06	7.50E-04

Table 2 illustrates the impact of using different values of k, with the error between $R_1^{(n,k)}$ and $R_1^{(n,\infty)}$ clearly decreasing as k increases. The simulations were run on a graph with $n = 10,000$ nodes. We also point out that although the accuracy of finitely many PageRank iterations improves as k gets larger, the

MSE of the tree approximation seems to plateau after a certain point. In order to obtain a higher level of precision we also need to increase the size of the graph (as suggested by Theorem 1).

Table 2. $n = 10000$, $\alpha = 2$, $\beta = 2.5$, $\lambda_1 = 1$, $c = 0.5$

k_n	$R_1^{(n,\infty)}$	$R_1^{(n,k_n)}$	$\hat{R}_1^{(n,k_n)}$	MSE for $R_1^{(n,k_n)}$	MSE for $\hat{R}_1^{(n,k_n)}$
2	0.908	0.933	0.928	7.1E-03	8.59E-03
4	0.929	0.933	0.933	1.5E-04	2.20E-04
6	0.908	0.909	0.910	5.4E-06	5.08E-05
8	0.883	0.884	0.884	8.8E-08	1.20E-06
10	0.948	0.949	0.950	7.6E-09	8.16E-05
15	0.932	0.932	0.932	7.9E-13	2.89E-05

Table 3 shows the same comparison as in Table 2, for fixed n, for different values of the damping factor c. As c gets larger, the approximations provided by both $R_1^{(n,k_n)}$ and $\hat{R}_1^{(n,k_n)}$ get worse due to the slower convergence of PageRank.

Our last numerical result shows how the distribution of PageRank on the TBT approximates the distribution of PageRank on the DCM. To illustrate this we generated a graph with $n = 100$ nodes and parameters $\alpha = 2$, $\beta = 2.5$, $\mu = 3$ and $c = 0.5$. We set the number of PageRank iterations (number of generations in the TBT) to be $k = 4$. We then computed the empirical CDFs of the PageRank of

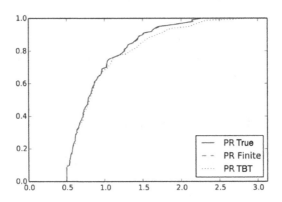

Fig. 2. The empirical distributions of PageRank on $\mathcal{G}_n$ (true and after finitely many iterations) and the empirical distribution of the PageRank of the root in the TBT

Table 3. $n = 10000$, $\alpha = 2$, $\beta = 2.5$, $\lambda_1 = 1$, $k_n = \lfloor \log n \rfloor = 9$

c	$R_1^{(n,\infty)}$	$R_1^{(n,k_n)}$	$\hat{R}_1^{(n,k_n)}$	MSE for $R_1^{(n,k_n)}$	MSE for $\hat{R}_1^{(n,k_n)}$
0.1	1.011	1.011	1.011	3.8E-22	3.33E-09
0.3	0.958	0.958	0.958	9.8E-13	1.91E-07
0.5	0.898	0.898	0.899	2.7E-08	2.63E-06
0.7	0.755	0.757	0.760	2.4E-05	2.03E-04
0.9	0.663	0.764	0.799	8.3E-02	1.25E-01

all nodes in the graph and that of the PageRank after only k iterations. We also generated the coupled TBT 1000 times based on the same graph; each time by randomly choosing some node i to be the root and computing $\hat{R}_i^{(n,k)}$ according to (7). Figure 2 plots the empirical CDF of PagerRank on $\mathcal{G}_n$, the empirical CDF of PageRank on $\mathcal{G}_n$ after only k iterations, and the empirical CDF of the PageRank of the 1000 root nodes after the same k iterations. We can see that the CDFs of PageRank on $\mathcal{G}_n$ after a finite number of iterations and that of the true PageRank on $\mathcal{G}_n$ are almost indistinguishable. The PageRank on the TBT also approximates this distribution quite well, especially considering that $n = 100$ is not particularly large.

References

1. Alsmeyer, G., Damek, E., Mentemeier, S.: Tails of fixed points of the two-sided smoothing transform. In: Springer Proceedings in Mathematics & Statistics: Random Matrices and Iterated Random Functions (2012)
2. Alsmeyer, G., Meiners, M.: Fixed points of the smoothing transform: Two-sided solutions. Probab. Theory Relat. Fields **155**(1–2), 165–199 (2013)
3. Avrachenkov, K., Lebedev, D.: PageRank of scale-free growing networks. Internet Mathematics **3**(2), 207–231 (2006)
4. Brin, S., Page, L.: The anatomy of a large-scale hypertextual Web search engine. Computer Networks and ISDN Systems **33**, 107–117 (1998)
5. Chen, N., Litvak, N., Olvera-Cravioto, M.: Ranking algorithms on directed configuration networks. ArXiv:1409.7443, pp. 1–39 (2014)
6. Chen, N., Olvera-Cravioto, M.: Directed random graphs with given degree distributions. Stochastic Systems **3**(1), 147–186 (2013)
7. Jelenković, P.R., Olvera-Cravioto, M.: Information ranking and power laws on trees. Adv. Appl. Prob. **42**(4), 1057–1093 (2010)
8. Jelenković, P.R., Olvera-Cravioto, M.: Implicit renewal theory and power tails on trees. Adv. Appl. Prob. **44**(2), 528–561 (2012)
9. Jelenković, P.R., Olvera-Cravioto, M.: Implicit renewal theory for trees with general weights. Stochastic Process. Appl. **122**(9), 3209–3238 (2012)
10. Langville, A.N., Meyer, C.D.: Google PageRank and beyond. Princeton University Press (2006)
11. Litvak, N., Scheinhardt, W.R.W., Volkovich, Y.: In-degree and PageRank: Why do they follow similar power laws? Internet Mathematics **4**(2), 175–198 (2007)
12. Olvera-Cravioto, M.: Tail behavior of solutions of linear recursions on trees. Stochastic Process. Appl. **122**(4), 1777–1807 (2012)
13. Pandurangan, G., Raghavan, P., Upfal, E.: Using PageRank to characterize web structure. In: Ibarra, O.H., Zhang, L. (eds.) COCOON 2002. LNCS, vol. 2387, pp. 330–339. Springer, Heidelberg (2002)
14. van der Hofstad, R.: Random graphs and complex networks (2009)
15. Volkovich, Y., Litvak, N.: Asymptotic analysis for personalized web search. Adv. Appl. Prob. **42**(2), 577–604 (2010)
16. Volkovich, Y., Litvak, N., Donato, D.: Determining factors behind the pagerank log-log plot. In: Proceedings of the 5th International Workshop on Algorithms and Models for the Web-graph, pp. 108–123 (2007)

Modelling of Trends in Twitter Using Retweet Graph Dynamics

Marijn ten Thij[1]([✉]), Tanneke Ouboter[2], Daniël Worm[2], Nelly Litvak[3],
Hans van den Berg[2,3], and Sandjai Bhulai[1]

[1] Faculty of Sciences, VU University Amsterdam, Amsterdam, The Netherlands
{m.c.ten.thij,s.bhulai}@vu.nl
[2] TNO, Delft, The Netherlands
{tanneke.ouboter,daniel.worm,j.l.vandenberg}@tno.nl
[3] Faculty of Electrical Engineering, Mathematics and Computer Science,
University of Twente, Enschede, The Netherlands
n.litvak@utwente.nl

Abstract. In this paper we model user behaviour in *Twitter* to capture the emergence of trending topics. For this purpose, we first extensively analyse tweet datasets of several different events. In particular, for these datasets, we construct and investigate the retweet graphs. We find that the retweet graph for a trending topic has a relatively dense largest connected component (LCC). Next, based on the insights obtained from the analyses of the datasets, we design a mathematical model that describes the evolution of a retweet graph by three main parameters. We then quantify, analytically and by simulation, the influence of the model parameters on the basic characteristics of the retweet graph, such as the density of edges and the size and density of the LCC. Finally, we put the model in practice, estimate its parameters and compare the resulting behavior of the model to our datasets.

Keywords: Retweet graph · Twitter · Graph dynamics · Random graph model

1 Introduction

Nowadays, social media play an important role in our society. The topics people discuss on-line are an image of what interests the community. Such trends may have various origins and consequences: from reaction to real-world events and naturally arising discussions to the trends manipulated e.g. by companies and organisations [14]. Trending topics on *Twitter* are 'ongoing' topics that become suddenly extremely popular[1]. In our study, we want to reveal differences in the retweet graph structure for different trends and model how these differences arise.

The work of Nelly Litvak is partially supported the EU-FET Open grant NADINE (288956).

[1] https://support.twitter.com/articles/101125-about-trending-topics

© Springer International Publishing Switzerland 2014
A. Bonato et al. (Eds.): WAW 2014, LNCS 8882, pp. 132–147, 2014.
DOI: 10.1007/978-3-319-13123-8_11

In *Twitter*[2] users can post messages that consist of a maximum of 140 characters. These messages are called tweets. One can "follow" a user in *Twitter*, which places their messages in the message display, called the timeline. Social ties are directed in *Twitter*, thus if user A follows user B, it does not imply that B follows A. People that "follow" a user are called "friends" of this user. We refer to the network of social ties in *Twitter* as the friend-follower network. Further, one can forward a tweet of a user, which is called a retweet.

There have been many studies on detecting different types of trends, for instance detecting emergencies [9], earthquakes [18], diseases [13] or important events in sports [11]. In many current studies into trend behaviour, the focus is mainly on content of the messages that are part of the trend, see e.g. [12]. Our work focuses instead on the underlying networks describing the social ties between users of *Twitter*. Specifically, we consider a graph of users, where an edge means that one of the users has retweeted a message of a different user.

In this study we use several datasets of tweets on multiple topics. First we analyse the datasets, described in Section 3, by constructing the retweet graphs and obtaining their properties as discussed in Section 4. Next, we design a mathematical model, presented in Section 5, that describes the growth of the retweet graph. The model involves two attachment mechanisms. The first mechanism is the preferential attachment mechanism that causes more popular messages to be retweeted with a higher probability. The second mechanism is the superstar mechanism which ensures that a user that starts a new discussion receives a finite fraction of all retweets in that discussion [2]. We quantify, analytically and with simulations, the influence of the model parameters on its basic characteristics, such as the density of edges, the size and the density of the largest connected component. In Section 6 we put the model in practice, estimate its parameters and compare it to our datasets. We find that what our model captures, is promising for describing the retweet graphs of trending topics. We close with conclusions and discussion in Section 7.

2   Related Work

The amount of literature regarding trend detection in *Twitter* is vast. The overview we provide here is by no means complete. Many studies have been performed to determine basic properties of the so-called "Twitterverse". Kwak et al. [10] analysed the follower distribution and found a non-power-law distribution with a short effective diameter and a low reciprocity. Furthermore they found that ranking by the number of followers and PageRank both induce similar rankings. They also report that *Twitter* is mainly used for News (85% of the content). Huberman et al. [8] found that the network of interactions within *Twitter* is not equal to the follower network, it is a lot smaller.

An important part of trending behaviour in social media is the way these trends progress through the network. Many studies have been performed on *Twitter* data. For instance, [3] studies the diffusion of news items in *Twitter* for

---

[2] www.twitter.com

several well-known news media and finds that these cascades follow a star-like structure. Also, [20] investigates the diffusion of information on *Twitter* using tweets on the Iranian election in 2009, and finds that cascades tend to be wide, not too deep and follow a power law-distribution in their size.

Bhamidi et al. [2] proposed and validated on the data a so-called superstar random graph model for a giant component of a retweet graph. Their model is based on the well-known preferential attachment idea, where users with many retweets have a higher chance to be retweeted [1], however, there is also a super-star node that receives a new retweet at each step with a positive probability. We build on this idea to develop our model for the progression of a trend through the *Twitter* network.

Another perspective on the diffusion of information in social media is obtained through analysing content of messages. For example, [17] finds that on *Twitter*, tags tend to travel to more distant parts of the network and URLs travel shorter distances. Romero et al. [16] analyse the spread mechanics of content through hashtag use and derive probabilities that users adopt a hashtag.

Classification of trends on *Twitter* has attracted considerable attention in the literature. Zubiaga et al. [21] derive four different types of trends, using 15 features to make their distinction. They distinguish trends triggered by news, current events, memes or commemorative tweets. Lehmann et al. [12] study different patterns of hashtag trends in *Twitter*. They also observe four different classes of hashtag trends. Rattanaritnont et al. [15] propose to distinguish topics based on four factors, which are cascade ratio, tweet ratio, time of tweet and patterns in topic-sensitive hashtags.

We extend the model of [2] by mathematically describing the growth of a complete retweet graph. Our proposed model has two more parameters that define the shape of the resulting graph, in particular, the size and the density of its largest connected component. To the best of our knowledge, this is the first attempt to classify trends using a random graph model rather than algorithmic techniques or machine learning. The advantage of this approach is that it gives insight in emergence of the trend, which, in turn, is important for understanding and predicting the potential impact of social media on real world events.

3   Datasets

We use datasets containing tweets that have been acquired either using the *Twitter* Streaming API[3] or the *Twitter* REST API[4]. Using the REST API one can obtain tweets or users from *Twitter*'s databases. The Streaming API filters tweets that *Twitter* parses during a day, for example, based on users, locations, hashtags, or keywords.

Most of the datasets used in this study were scraped by RTreporter, a company that uses an incoming stream of Dutch tweets to detect news for news agencies in the Netherlands. These tweets are scraped based on keywords, using

---

[3] https://dev.twitter.com/docs/streaming-apis
[4] https://dev.twitter.com/docs/api/1.1

the Streaming API. For this research, we selected several events that happened in the period of data collection, based on the wikipedia overviews of 2013 and 2014[5]. We have also used two datasets scraped by TNO - Netherlands Organisation for Applied Scientific Research. The *Project X* dataset contains tweets related to large riots in Haren, the Netherlands. This dataset is acquired by *Twitcident*[6]. For this study, we have filtered this dataset on two most important hashtags: *#projectx* and *#projectxharen*. The *Turkish-Kurdish* dataset is described in more detail in Bouma et al. [4]. A complete overview of the datasets, including the events and the keywords, is given in Table 1. The size and the timespans for each dataset are given in Table 2.

Table 1. Datasets: events and keywords (some keywords are in Dutch)

	dataset	keywords
PX	Project X Haren	projectx, projectxharen
TK	Demonstrations in Amsterdam related to the Turkish-Kurdish conflict	koerden, turken, rellen, museumplein, amsterdam
WCS	World cup speedskating single distanced 2013	wkafstanden, sochi, sotsji
W-A	Crowning of His Majesty King Willem-Alexander in the Netherlands	troonswisseling, troon, Willem-Alexander, Wim-Lex, Beatrix, koning, koningin
ESF	Eurovision Song Festival	esf, Eurovisie Songfestival, ESF, songfestival, eurovisie
CL	Champions Leage final 2013	Bayern Munchen, Borussia Dortmund, dorbay, borussia, bayern, borbay, CL
Morsi	Morsi deposited as Egyption president	Morsi, afgezet, Egypte
Train	Train crash in Santiago, Spain	Treincrash, treincrash, Santiago, Spanje, Santiago de Compostella, trein
Heat	Heat wave in the Netherlands	hittegolf, Nederland
Damascus	Sarin attack in Damascus	Sarin, Damascus, Syrië, syrië
Peshawar	Bombing in Peshawar	Peshawar, kerk, zelfmoordaanslag, Pakistan
Hawk	Hawk spotted in the Netherlands	sperweruil, Zwolle
Pile-up	Multiple pile-ups in Belgium on the A19	A19, Ieper, Kortrijk, kettingbotsing
Schumi	Michael Schumachar has a skiing accident	Michael Schumacher, ski-ongeval
UKR	Rebellion in Ukrain	Azarov, Euromaidan, Euromajdan, Oekraïne, opstand
NAM	Treaty between NAM and Dutch government	Loppersum, gasakkoord, NAM, Groningen
WCD	Michael van Gerwen wins PDC WC Darts	van Gerwen, PDC, WK Darts
NSS	Nuclear Security Summit 2014	NSS2014, NSS, Nuclear Security Summit 2014, Den Haag
MH730	Flight MH730 disappears	MH730, Malaysia Airlines
Crimea	Crimea referendum for independance	Krim, referendum, onafhankelijkheid
Kingsday	First Kingsday in the Netherlands	koningsdag, kingsday, koningsdag
Volkert	Volkert van der Graaf released from prison	Volkert, volkertvandergraaf, Volkert van der Graaf

For each dataset we have observed there is at least one large peak in the progression of the number of tweets. For example, Figure 1 shows such peak in *Twitter* activity for the *Project X* dataset.

When a retweet is placed on *Twitter*, the Streaming API returns the retweet together with the message that has been retweeted. We use this information to construct the retweet trees of every message and the user IDs for each posted message. The tweet and graph analysis is done using *Python* and its modules

---

[5] http://nl.wikipedia.org/wiki/2014 & http://nl.wikipedia.org/wiki/2013
[6] www.twitcident.com

Table 2. Characteristics of the datasets

dataset	year	first tweet	last tweet	# tweets	# retweets
PX	2012	Sep 17 09:37:18	Sep 26 02:31:15	31,144	15,357
TK	2011	Oct 19 14:03:23	Oct 27 08:42:18	6,099	999
WCS	2013	Mar 21 09:19:06	Mar 25 08:45:50	2,182	311
W-A	2013	Apr 27 22:59:59	May 02 22:59:25	352,157	88,594
ESF	2013	May 13 23:00:08	May 18 22:59:59	318,652	82,968
CL	2013	May 22 23:00:04	May 26 22:59:54	163,612	54,471
Morsi	2013	Jun 30 23:00:00	Jul 04 22:59:23	40,737	13,098
Train	2013	Jul 23 23:00:02	Jul 30 22:59:41	113,375	26,534
Heat	2013	Jul 10 19:44:35	Jul 29 22:59:58	173,286	42,835
Damascus	2013	Aug 20 23:01:57	Aug 31 22:59:54	39,377	11,492
Peshawar	2013	Sep 21 23:00:00	Sep 24 22:59:59	18,242	5,323
Hawk	2013	Nov 11 23:00:07	Nov 30 22:58:59	54,970	19,817
Pile-up	2013	Dec 02 23:00:15	Dec 04 22:59:57	6,157	2,254
Schumi	2013-14	Dec 29 02:43:16	Jan 01 22:54:50	13,011	5,661
UKR	2014	Jan 26 23:00:36	Jan 31 22:57:12	4,249	1,724
NAM	2014	Jan 16 23:00:22	Jan 20 22:59:49	41,486	14,699
WCD	2013-14	Dec 31 23:03:48	Jan 02 22:59:05	15,268	5,900
NSS	2014	Mar 23 23:00:06	Mar 24 22:59:56	29,175	13,042
MH730	2014	Mar 08 00:18:32	Mar 28 22:40:44	36,765	17,940
Crimea	2014	Mar 13 23:02:22	Mar 17 22:59:57	18,750	5,881
Kingsday	2014	Apr 26 23:00:00	Apr 29 22:53:00	7,576	2,144
Volkert	2014	Apr 30 23:08:14	May 04 22:57:06	9,659	4,214

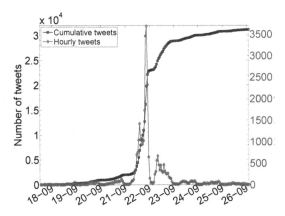

Fig. 1. *Project X* Number of tweets and cumulative number of tweets per hour

Tweepy[7] and *NetworkX*[8]. In this paper, we investigate the dynamics of retweet graphs with the goal to predict peaks in *Twitter* activity and classify the nature of trends.

4   Retweet Graphs

Our main object of study is the retweet graph $G = (V, E)$, which is a graph of users that have participated in the discussion on a specific topic. A directed

---

[7] http://www.tweepy.org/
[8] http://networkx.github.io/

edge $e = (u, v)$ indicates that user v has retweeted a tweet of u. We observe
the retweet graph at the time instances $t = 0, 1, 2, \ldots$, where either a new node
or a new edge was added to the graph, and we denote by $G_t = (V_t, E_t)$ the
retweet graph at time t. As usual, the out- (in-) degree of node u is the number
of directed edges with source (destination) in u. In what follows, we model and
analyse the properties of G_t. For every new message initiated by a new user u
a tree T_u is formed. Then, $\mathcal{T}_t$ denotes the forest of message trees. Note that in
our model a new message from an already existing user u (that is, $u \in \mathcal{T}_t$) does
not initiate a new message tree. We define $|\mathcal{T}_t|$ as the number of new users that
have started a message tree up to time t.

After analyzing multiple characteristics of the retweet graphs for every hour
of their progression, we found that the size of the largest (weakly) connected
component (LCC) and its density are the most informative characteristics for
predicting the peak in *Twitter*. In Figure 2 we show the development of these
characteristics in the *Project X* dataset. One day before the actual event, we
observe a very interesting phenomenon in the development of the edge density
of the LCC in Figure 2a. Namely, at some point the edge density of the LCC
exceeds 1 (indicated by the dash-dotted gray lines), i.e. there is more than one
retweet per user on average. We shall refer to this as the *densification* (or dens.)
of the LCC. Furthermore, the relative size of the LCC increases from 18% to
25% as well, see Figure 2b.

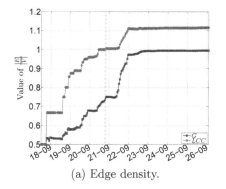 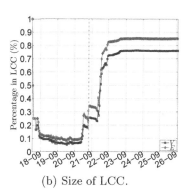

(a) Edge density.   (b) Size of LCC.

Fig. 2. Progression for the edge density (a) and the size of the LCC (b) in the *Project X* dataset

We have observed a densification of the LCC in each dataset that we have
studied. Indeed, when the LCC grows its density must become at least one (each
node is added to the LCC together with at least one edge). However, we have also
observed that in each dataset the densification occurs before the main peak, but
the scale of densification is different. For example, in the *Project X* dataset the
densification already occurs one day before the peak activity. Plausibly, in this
discussion, that ended up in riots, a group of people was actively participating
before the event. On the other hand, in the *WCS* dataset, which tweets about
an ongoing sport event, the densication of the LCC occurs during the largest

peak. This is the third peak in the progression. Hence, our experiments suggest that the time of densification has predictive value for trend progression and classification. See Table 3 for the density of the LCC in each dataset at the end of the progression.

5   Model

Our goal is to design a model that captures the development of trending behaviour. In particular, we need to capture the phenomenon that disjoint components of the retweet graph join together forming the largest component, of which the density of edges may become larger than one. To this end, we employ the superstar model of Bhamidi et al. [2] for modelling distinct components of the retweet graph, and add the mechanism for new components to arrive and the existing components to merge. For the sake of simplicity of the model we neglect the friend-follower network of *Twitter*. Note that in *Twitter* every user can retweet any message sent by any public user, which supports our simplification.

At the start of the progression, we have the graph G_0. In the analysis of this section, we assume that G_0 consists of a single node. Note that in reality, this does not need to be the case: any directed graph can be used as an input graph G_0. In fact, in Section 6 we start with the actual retweet graph at a given point in time, and then use the model to build the graph further to its final size.

We consider the evolution of the retweet graph in time $(G_t)_{t \geq 0}$. We use a subscript t to indicate G_t and related notions at time t. We omit the index t when referring to the graph at the end of the progression.

Recall that G_t is a graph of *users*, and an edge (u, v) means that v has retweeted a tweet of u. We consider time instances $t = 1, 2, \ldots$ when either a new node or a new edge is added to the graph G_{t-1}. We distinguish three types of changes in the retweet graph:

- $T1$: a new user u has posted a new message on the topic, node u is added to G_{t-1};
- $T2$: a new user v has retweeted an existing user u, node v and edge (u, v) are added to G_{t-1};
- $T3$: an existing user v has retweeted another existing user u, edge (u, v) is added to G_{t-1}.

The initial node is equivalent to a $T1$ arrival at time $t = 0$. Assume that each change in G_t at $t = 1, 2, \ldots$ is $T1$ with probability $\lambda/(1 + \lambda)$, independently of the past. Also, assume that a new edge (retweet) is coming from a new user with probability p. Then the probabilities of $T1$, $T2$ and $T3$ arrivals are, respectively $\frac{\lambda}{\lambda+1}, \frac{p}{\lambda+1}, \frac{1-p}{\lambda+1}$. The parameter p is governing the process of components merging together, while λ is governing the arrival of new components in the graph.

For both $T2$ and $T3$ arrivals we define the same mechanism for choosing the source of the new edge (u, v) as follows.

Let $u_0, u_1, \ldots$ be the users that have been added to the graph as $T1$ arrivals, where u_0 is the initial node. Denote by $T_{i,t}$ the subgraph of G_t that includes u_i

and all users that have retweeted the message of u_i in the interval $(0, t]$. We call such a subgraph a message tree with root u_i. We assume that the probability that a $T2$ or $T3$ arrival at time t will attach an edge to one of the nodes in $T_{i,t-1}$ with probability $p_{T_{i,t-1}}$, proportional to the size of the message tree:

$$p_{T_{i,t-1}} = \frac{|T_{i,t-1}|}{\sum_{T_{j,t-1} \subset T_{t-1}} |T_{j,t-1}|}.$$

This creates a preferential attachment mechanism in the formation of the message trees. Next, a node in the selected message tree $T_{i,t-1}$ is chosen as the source node following the superstar attachment scheme [2]: with probability q, the new retweet is attached to u_i, and with probability $1 - q$, the new retweet is attached to any other vertex, proportional to the preferential attachment function of the node, that we choose to be the number of children of the node plus one.

Thus we employ the superstar-model, which was suggested in [2] for modelling the largest connected component of the retweet graph on a given topic, in order to describe a progression mechanism for a single retweet tree. Our extensions compared to [2] are that we allow new message trees to appear ($T1$ arrivals), and that different message trees may either remain disconnected or get connected by a $T3$ arrival.

For a $T3$ arrival, the target of the new edge (u, v) is chosen uniformly at random from V_{t-1}, with the exception of the earlier chosen source node u, to prevent self-loops. That is, any user is equally likely to retweet a message from another existing user.

Note that, in our setting, it is easy to introduce a different superstar parameter q_{T_i} for every message tree T_i. This way one could easily implement specific properties of the user that starts the message tree, e.g. his/her number of followers. For the sake of simplicity, we choose the same value of q for all message trees. Also note that we do not include tweets and retweets that do not result in new nodes or edges in a retweet graph. This could be done, for example, by introducing dynamic weights of vertices and edges, that increase with new tweets and retweets. Here we consider only an unweighted model.

5.1   Growth of the Graph

The average degree, or edge density, is one of the aspects through which we give insight to the growth of the graph. The essential properties of this characteristic are presented in Theorem 1. The proof is given in the Appendix.

Theorem 1. *Let τ_n be the time when node n is added to the graph. Then*

$$\mathbb{E}\left[\frac{|E_{\tau_n}|}{|V_{\tau_n}|}\right] = \frac{1}{\lambda + p} - \frac{1}{n(\lambda + p)}, \tag{1}$$

$$var\left(\frac{|E_{\tau_n}|}{|V_{\tau_n}|}\right) = \frac{(n-1)(\lambda + 1 - p)}{n^2(\lambda + p)^2}. \tag{2}$$

Note that the variance of the average degree in (2) converges to zero as $n \to \infty$ at rate $\frac{1}{n}$.

The next theorem studies the observed ratio between $T2$ and $T3$ arrivals (new edges) and $T1$ arrivals (new nodes with a new message). As we see from the theorem, this ratio can be used for estimating the parameter λ. The proof is given in the Appendix.

Theorem 2. *Let $G_t = (V_t, E_t)$ be the retweet graph at time t, let $\mathcal{T}_t$ be the set of all message trees in G_t. Then*

$$\mathbb{E}\left[\frac{|E_t|}{|\mathcal{T}_t|}\right] = \lambda^{-1} \cdot \left(1 - \left(\frac{1}{\lambda+1}\right)^t\right), \tag{3}$$

$$\lim_{t\to\infty} \frac{\lambda^3 t}{(1+\lambda)^2} var\left(\frac{|E_t|}{|\mathcal{T}_t|}\right) = 1, \tag{4}$$

Furthermore,

$$\frac{\lambda^{3/2}\sqrt{t}}{\lambda+1}\left(\frac{|E_t|}{|\mathcal{T}_t|} - \frac{1}{\lambda}\right) \xrightarrow{D} Z, \tag{5}$$

where Z is a standard normal $N(0,1)$ random variable, and $\xrightarrow{D}$ denotes convergence in distribution.

Note that, as expected from the definition of λ,

$$\lim_{t\to\infty} \mathbb{E}\left[\frac{|E_t|}{|\mathcal{T}_t|}\right] = \lambda^{-1}. \tag{6}$$

This will be used in Section 6 for estimating λ.

5.2   Component Size Distribution

In the following, we assume that G_t consists of m connected components $(C_1, C_2, \ldots, C_m)$ with known respective sizes $(|C_1|, \ldots, |C_m|)$. We aim to derive expressions for the distribution of the component sizes in G_{t+1}.

Lemma 3. *The distribution of the sizes of the components of G_{t+1}, given G_t is as follows,*

$$\begin{cases} |C_1|, \ldots, |C_i|, |C_j|, \ldots, |C_m|, 1 & w.p. \ \frac{\lambda}{\lambda+1} \\ |C_1|, \ldots, |C_i|+1, |C_j|, \ldots, |C_m| & w.p. \ \frac{p}{\lambda+1} \cdot \frac{|C_i|}{|V|} \\ |C_1|, \ldots, |C_i|+|C_j|, \ldots, |C_m| & w.p. \ \frac{1-p}{\lambda+1} \cdot \frac{2 \cdot |C_i| \cdot |C_j|}{|V|^2 - |V|} \\ |C_1|, \ldots, |C_i|, |C_j|, \ldots, |C_m| & w.p. \ \frac{1-p}{\lambda+1} \cdot \frac{\sum_{k=1}^m |C_k|^2 - |C_k|}{|V|^2 - |V|} \end{cases} \tag{7}$$

The proof of Lemma 3 is given in the Appendix. Lemma 3 provides a recursion for computing the distribution of component sizes. However, the computations are highly demanding if not infeasible. Also, deriving an exact expression of the distribution of the component sizes at time t is very cumbersome because they are hard and they strongly depend on the events that occurred at $t = 0, \ldots, t - 1$. Note that if $p = 1$, there is a direct correspondence between our model and the infinite generalized Pólya process [5]. However, this case is uninformative as there are no $T3$ arrivals. Therefore, in the next section we resort to simulations to investigate the sensitivity of the graph characteristics to the model parameters.

5.3   Influence of q, p and λ

We analyze the influence of the model parameters λ, p and q on the characteristics of the resulting graph numerically using simulations. To this end, we fix two out of three parameters and execute multiple simulation runs of the model, varying the values for the third parameter. We start simulations with graph G_0, consisting of one node. We perform 50 simulation runs for every parameter setting and obtain the average values over the individual runs for given parameters.

Parameter q affects the degree distribution [2] and the overall structure of the graph. If $q = 0$, then the graph contains less nodes that have many retweets. If $q = 1$ each edge is connected to a superstar, and the graph consists of star-like sub graphs, some of which are connected to each other. In the *Project X* dataset, which is our main case study, $q \approx 0.9$ results in a degree distribution that closely approximates the data. Since degree distributions are not in the scope of this paper, we omit these results for brevity.

We compare the results for two measures that produced especially important characteristics of the *Project X* dataset: $\frac{|E_{\text{LCC}}|}{|V_{\text{LCC}}|}$ and $\frac{|V_{\text{LCC}}|}{|V|}$. These characteristics do not depend on q. In simulations, we set $t = 1,000$, $q = 0.9$ and vary the values for p and λ. the results are give in Figure 3.

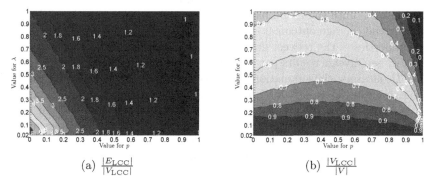

(a) $\frac{|E_{\text{LCC}}|}{|V_{\text{LCC}}|}$ (b) $\frac{|V_{\text{LCC}}|}{|V|}$

Fig. 3. Numerical results for the model using $q = 0.9$ and $t = 1,000$

We see that the edge density in the LCC in Figure 3a decreases with λ and p. Note that according to (1), $|E|/|V|$ is well approximated by $1/(\lambda + p)$ when λ or p are large enough. The edge density in LCC shows a similar pattern, but it is slightly higher than in the whole graph. When λ and p are small, there are many $T3$ arrivals, and new nodes are not added frequently enough. This results in an unexpected non-monotonic behaviour of the edge density near the origin. For the fraction of nodes in the LCC, depicted in Figure 3b, we see that the parameter λ is most influential. The parameter p is of considerable influence only when it is large.

6   The Model in Practice

In this section we obtain parameter estimators for our model and compare the model to the datasets discussed in Section 3.

Using Theorem 2, we know that $\frac{|E_t|}{|T_t|}$ converges to λ^{-1} as $t \to \infty$. Thus, we suggest the following estimator for λ at time $t > 0$:

$$\hat{\lambda}_t = \frac{|T_t|}{|E_t|}. \tag{8}$$

Second, we derive an expression for $\hat{p}_t$ using (1) and substituting (8) for λ:

$$\hat{p}_t = \frac{|V_t| - |T_t| - 1}{|E_t|}. \tag{9}$$

Since the *Twitter* API only gives back the original message of a retweet and not the level in the progression tree of that retweet, we can not determine q easily from the data. Since this parameter does not have a large influence on the outcomes of the simulations, we choose this parameter to be 0.9 for all datasets.

Notice that we can obtain the numbers ($|E_t|$, $|T_t|$ and $|V_t|$) directly from a given retweet graph for each $t = 1, 2, \ldots$. The computed estimators for our datasets are displayed in Table 3.

Next, we compare 50 simulations of the datasets from the point of densification of the LCC until the graph has reached the same size as the actual dataset. We display the average outcomes of these simulations and compare them to the actual properties of the retweet graphs of each dataset in Table 3.

Here we see diverse results per dataset in the simulations. For the *CL, Morsi* and *WCD* datasets, the simulations are very similar to the actual progressions. However, for some datasets, for instance the *ESF* dataset, simulations are far off. In general, the model predicts the density of the LCC quite well for many datasets, but tends to overestimate the size of the LCC. We notice that current random graph models for networks usually capture one or two essential features, such as degree distribution, self-similarity, clustering coefficient or diameter. Our model captures both degree distribution and, in many cases, the density of the LCC. It seems that our model performs better on the datasets that have a

Table 3. Estimated parameter values using complete dataset, simulation and progression properties

dataset	$\hat{\lambda}$	$\hat{p}$	actual progression $\frac{\|V_{LCC}\|}{\|V\|}$	$\frac{\|E\|}{\|V\|}$	$\frac{\|E_{LCC}\|}{\|V_{LCC}\|}$	simulations (starting at dens.) $\frac{\|V_{LCC}\|}{\|V\|}$	$\frac{\|E\|}{\|V\|}$	$\frac{\|E_{LCC}\|}{\|V_{LCC}\|}$
PX	.23	.78	.76	1.00	1.12	.54	.75	1.08
TK	.42	.85	.25	.79	1.00	.54	.74	1.08
WCS	.49	.73	.20	.81	.99	.49	.95	1.90
W-A	.41	.52	.67	1.07	1.30	.40	.62	1.41
ESF	.38	.43	.73	1.24	1.48	.45	.69	1.42
CL	.40	.72	.44	.90	1.22	.46	.66	1.16
Morsi	.60	.55	.39	.87	1.20	.47	.67	1.17
Train	.54	.78	.28	.76	1.04	.50	.70	1.17
Heat	.42	.59	.60	.99	1.23	.41	.72	1.68
Damascus	.58	.51	.46	.92	1.24	.44	.65	1.30
Peshawar	.54	.68	.31	.82	1.18	.53	.75	1.25
Hawk	.38	.38	.82	1.31	1.45	.49	.76	1.43
Pile-up	.33	.64	.65	1.03	1.24	.58	.93	1.54
Schumi	.38	.83	.33	.82	1.08	.56	.77	1.07
UKR	.72	.37	.53	.91	1.12	.50	.75	1.38
NAM	.44	.48	.50	1.09	1.51	.45	.72	1.51
WCD	.26	.81	.66	.94	1.10	.64	.83	1.07
NSS	.26	.62	.79	1.13	1.26	.23	.35	1.21
MH730	.33	.52	.15	1.18	1.00	.56	.76	1.09
Crimea	.44	.63	.51	.93	1.19	.52	.72	1.12
Kingsday	.47	.92	.07	.72	1.11	.47	.67	1.15
Volkert	.29	.55	.79	1.18	1.31	.64	.87	1.22

singular peak rather than a series of peaks. We have observed on the data that each peak activity has a large impact on the parameters estimation. We will strive to adopt the model for incorporating different rules for activity during peaks, and improving results on the size of the LCC.

7 Conclusion and Discussion

We have found that our model performs well in modelling the retweet graph for tweets regarding a singular topic. However, there is a room for improvement when the dataset covers a prolonged discussion with users activity fluctuating over time.

A possible extension of the present work is incorporating more explicitly the time aspect into our model. We could for example add the notion of 'novelty', like Gómez et al. in [6], taking into account that e.g. the retweet probability for a user may decrease the longer he/she remains silent after having received a tweet. But also other model parameters may be assumed to vary over time. In addition, we propose to analyse the clustering coefficient of a node in the network model and, in particular, to investigate how it evolves over time. This measure (see [19]) provides more detailed insight in how the graph becomes denser, making it possible to distinguish between local and global density.

Appendix

A1. Proof of Theorem 1

Proof. The proof is based on the fact that the total number of edges $|E_{\tau_n}|$ equals a total number of the $T2$ and $T3$ arrivals on $(0, \tau_n]$. By definition, $(0, \tau_n]$ contains exactly $(n-1)$ of $T1$ or $T2$ arrivals, hence, the number of $T2$ arrivals has a Binomial distribution with number of trials equal to $(n-1)$, and success probability $P(T2 \mid T1 \text{ or } T2) = \frac{p}{\lambda+p}$. Next, the number of $T3$ arrivals on $[\tau_i, \tau_{i+1})$, where $i = 1, \ldots, n-1$, has a shifted geometric distribution, namely, the probability of k $T3$ arrivals on $[\tau_i, \tau_{i+1})$ is

$$\left(1 - \frac{1-p}{\lambda+1}\right)\left(\frac{1-p}{\lambda+1}\right)^k, \quad k = 0, 1, \ldots.$$

Observe that there have been $n-1$ of these transitions from 1 node to n. Hence, the number of $T3$ arrivals on $(0, \tau_n]$ is the sum of $(n-1)$ i.i.d. Geometric random variables with mean $\frac{1-p}{\lambda+p}$. Summarizing the above, we obtain (1). For (2) we also need to observe that the number of $T2$ and $T3$ arrivals on $[0, \tau_n]$ are independent.

A2. Proof of Theorem 2

Proof. Let X_t be the number of $T2$ and $T3$ arrivals by time t. Note that $|E_t| = X_t$, and $|T_t| = t - X_t + 1$, which is the number of T_1 arrivals on $[0, t]$, since the first node at time $t = 0$ is by definition a $T1$ arrival. Note that X_t has a binomial distribution with parameters t and $\mathbb{P}(T2 \text{ arrival}) + \mathbb{P}(T3 \text{ arrival}) = \frac{1}{\lambda+1}$. Furthermore, the number of $T1$ arrivals is $t - X_t + 1$ since the first node at time $t = 0$ is by definition a $T1$ arrival. Hence,

$$\mathbb{E}\left[\frac{|E_t|}{|T_t|}\right] = \sum_{i=1}^{t} \frac{i}{t-i+1}\binom{t}{i}\left(\frac{1}{\lambda+1}\right)^i\left(\frac{\lambda}{\lambda+1}\right)^{t-i}$$

$$= \frac{1}{\lambda} \cdot \sum_{i=1}^{t}\binom{t}{i-1}\left(\frac{1}{\lambda+1}\right)^{i-1}\left(\frac{\lambda}{\lambda+1}\right)^{t-i+1},$$

which proves (3). Next, we write

$$\mathbb{E}\left[\left(\frac{|E_t|}{|T_t|}\right)^2\right] = \sum_{i=0}^{t}\left(\frac{i}{t-i+1}\right)^2\binom{t}{i}\left(\frac{1}{\lambda+1}\right)^i\left(\frac{\lambda}{\lambda+1}\right)^{t-i}$$

$$= \frac{1}{\lambda} \cdot \sum_{i=1}^{t}\frac{i}{t-i+1}\binom{t}{i-1}\left(\frac{1}{\lambda+1}\right)^{i-1}\left(\frac{\lambda}{\lambda+1}\right)^{t-i+1}$$

$$= \frac{1}{\lambda}\mathbb{E}\left[\frac{t+1}{t-X_t}\mathbb{1}_{\{X_t \leq t-1\}}\right] - \frac{1}{\lambda}\left(1 - \left(\frac{1}{1+\lambda}\right)^t\right), \tag{10}$$

where $\mathbb{1}_{\{A\}}$ is an indicator of event A. Denoting

$$Z_t = \frac{X_t - \mathbb{E}[X_t]}{\sqrt{var(X_t)}} = \frac{(\lambda+1)X_t - t}{\sqrt{\lambda t}}, \tag{11}$$

we further write

$$\mathbb{E}\left[\frac{t+1}{t-X_t}\mathbb{1}_{\{X_t \leq t-1\}}\right] = \mathbb{E}\left[\frac{(t+1)(\lambda+1)}{\lambda t(1-\frac{Z_t}{\sqrt{\lambda t}})}\mathbb{1}_{\{Z_t \leq \sqrt{\lambda t}-\frac{\lambda+1}{\sqrt{\lambda t}}\}}\right]. \tag{12}$$

We now split the indicator above as follows:

$$\mathbb{1}_{\{Z_t \leq -\sqrt{\lambda t}\}} + \mathbb{1}_{\{-\sqrt{\lambda t} < Z_t < \sqrt{\lambda t}/2\}} + \mathbb{1}_{\{\sqrt{\lambda t}/2 \leq Z_t \leq \sqrt{\lambda t}-\frac{\lambda+1}{\sqrt{\lambda t}}\}}. \tag{13}$$

For the first and the third term we use the Chernoff bound:

$$\mathbb{E}\left[\frac{1}{1-\frac{Z_t}{\sqrt{\lambda t}}}\mathbb{1}_{\{Z_t \leq -\sqrt{\lambda t}\}}\right] \leq 2e^{-\lambda t/4}, \tag{14}$$

$$\mathbb{E}\left[\frac{1}{1-\frac{Z_t}{\sqrt{\lambda t}}}\mathbb{1}_{\{\sqrt{\lambda t}/2 \leq Z_t \leq \sqrt{\lambda t}-\frac{\lambda+1}{\sqrt{\lambda t}}\}}\right] \leq \frac{\sqrt{\lambda t}}{\lambda+1}2e^{-\lambda t/16}, \tag{15}$$

and notice that both expressions above converge to zero faster than $1/t$. For the second case, note first that $\mathbb{E}[Z_t] = 0$ and hence it follows from (11) and (13)–(15) that, as $t \to \infty$,

$$\mathbb{E}\left[Z_t \mathbb{1}_{\{-\sqrt{\lambda t} < Z_t < \sqrt{\lambda t}/2\}}\right] = o\left(\frac{1}{t}\right).$$

Then we use the Taylor expansion to obtain:

$$\left|\mathbb{E}\left[\frac{1}{1-\frac{Z_t}{\sqrt{\lambda t}}}\mathbb{1}_{\{-\sqrt{\lambda t} < Z_t < \sqrt{\lambda t}/2\}}\right] - 1\right|$$
$$\leq \mathbb{E}\left[\frac{Z_t^2}{\lambda t}\right] + 2\mathbb{E}\left[\frac{|Z_t|^3}{(\lambda t)^{3/2}}\right] + o\left(\frac{1}{t}\right), \tag{16}$$

as $t \to \infty$. By the central limit theorem, $Z_t \xrightarrow{D} Z$ as $t \to \infty$. Furthermore, for $r > 0$, the convergence of moments holds [7]: $\lim_{t\to\infty}\mathbb{E}[|Z_t|^r] = \mathbb{E}[|Z|^r]$. In particular, in (16), $\mathbb{E}[|Z_t|^3]$ converges to a constant, and $\mathbb{E}[Z_t^2]$ converges to 1 as $t \to \infty$. Thus, using (10)–(12) and (3) we write

$$var\left(\frac{|E_t|}{|T_t|}\right) = \mathbb{E}\left[\left(\frac{|E_t|}{|T_t|}\right)^2\right] - \left(\mathbb{E}\left[\frac{|E_t|}{|T_t|}\right]\right)^2$$
$$= \mathbb{E}\left[\frac{(t+1)(\lambda+1)}{\lambda t(1-\frac{Z_t}{\sqrt{\lambda t}})}\mathbb{1}_{\{Z_t \leq \sqrt{\lambda t}-\frac{\lambda+1}{\sqrt{\lambda t}}\}}\right] - \frac{1}{\lambda} - \frac{1}{\lambda^2} + o\left(\frac{1}{t}\right).$$

Now, subsequently using (13) – (16), we get

$$var\left(\frac{|E_t|}{|T_t|}\right) = \frac{1}{\lambda}\frac{(t+1)(\lambda+1)}{\lambda t}\left(1 + \frac{1}{\lambda t} + o\left(\frac{1}{t}\right)\right)$$
$$-\frac{1}{\lambda} - \frac{1}{\lambda^2} + o\left(\frac{1}{t}\right),$$

which results in (4). Statement (5) is proved along similar lines: we apply the expansion directly to the random variable

$$\frac{X_t}{t - X_t + 1} = \frac{(t+1)(\lambda+1)}{(\lambda t + \lambda + 1)(1 - Z_t\frac{\sqrt{\lambda t}}{\lambda t + \lambda + 1})}\mathbb{1}_{\{Z_t \le \sqrt{\lambda t}\}} - 1,$$

and then use the Chernoff bounds and the CLT to obtain the result.

A3. Proof of Lemma 3

Proof. Assume the arrival at time $t+1$ is of type $T1$. This occurs w.p. $\frac{\lambda}{\lambda+1}$, and then a new component consisting of size one is created in G_{t+1}, corresponding to the first case in (3).

Next, consider a $T2$ arrival, which occurs w.p. $\frac{p}{\lambda+1}$. We now add a node to an existing component C_i w.p. $\frac{|C_i|}{|V|}$. Thus the probability that we add the new node to C_i is $\frac{p}{\lambda+1} \cdot \frac{|C_i|}{|V|}$.

Last, we consider a $T3$ arrival. In this case we have two options. The new edge can either join two components, or join two nodes that are already in one component. For the first case, we derive the probability that C_i and C_j join as

$$\mathbb{P}\left(C_i \text{ and } C_j \text{ merge}\right) = \frac{1-p}{\lambda+1} \cdot \frac{2 \cdot |C_i| \cdot |C_j|}{|V|^2 - |V|}.$$

Then for the second case, the number of ways a T3 arrival links two nodes that are already connected in a component, say C_i, is $|C_i|(|C_i| - 1)$. Therefore with probability $\frac{\sum_{k=1}^{m}|C_k|^2 - |C_k|}{|V|^2 - |V|}$ the component size does not change.

References

1. Barabási, A.-L., Albert, R.: Emergence of scaling in random networks. Science **286**(5439), 509–512 (1999)
2. Bhamidi, S., Steele, J.M., Zaman, T.: Twitter event networks and the superstar model, arXiv preprint ar Xiv:1211.3090 (2012)
3. Bhattacharya, D., Ram, S.: Sharing news articles using 140 characters: A diffusion analysis on Twitter. In: 2012 IEEE/ACM International Conference on Advances in Social Networks Analysis and Mining, ASONAM, pp. 966–971. IEEE (2012)
4. Bouma, H., Rajadell, O., Worm, D., Versloot, C., Wedemeijer, H.: On the early detection of threats in the real world based on open-source information on the Internet. In: International Conference on Information Technologies and Security, ITSEC (2012)

5. Chung, F., Handjani, S., Jungreis, D.: Generalizations of Polya's urn problem. Annals of Combinatorics **7**(2), 141–153 (2003)
6. Gómez, V., Kappen, H.J., Litvak, N., Kaltenbrunner, A.: A likelihood-based framework for the analysis of discussion threads. World Wide Web, 1–31 (2012)
7. Hall, P.: On the rate of convergence of moments in the central limit theorem for lattice distributions. Transactions of the American Mathematical Society **278**(1), 169–181 (1983)
8. Huberman, B., Romero, D., Wu, F.: Social networks that matter: Twitter under the microscope. Available at SSRN 1313405 (2008)
9. Klein, B., Laiseca, X., Casado-Mansilla, D., López-de-Ipiña, D., Nespral, A.P.: Detection and extracting of emergency knowledge from twitter streams. In: Bravo, J., López-de-Ipiña, D., Moya, F. (eds.) UCAmI 2012. LNCS, vol. 7656, pp. 462–469. Springer, Heidelberg (2012)
10. Kwak, H., Lee, C., Park, H., Moon, S.: What is Twitter, a social network or a news media? In: Proceedings of the 19th International Conference on World Wide Web, pp. 591–600. ACM (2010)
11. Lanagan, J., Smeaton, A.F.: Using Twitter to detect and tag important events in live sports, pp. 542–545. AAAI (2011)
12. Lehmann, J., Gonçalves, B., Ramasco, J.J., Cattuto, C.: Dynamical classes of collective attention in Twitter. In: Proceedings of the 21st International Conference on World Wide Web, pp. 251–260. ACM (2012)
13. Paul, M.J., Dredze, M.: You are what you tweet: Analyzing Twitter for public health. In: Fifth International AAAI Conference on Weblogs and Social Media, ICWSM 2011 (2011)
14. Ratkiewicz, J., Conover, M., Meiss, M., Gonçalves, B., Patil, S., Flammini, A., Menczer, F.: Truthy: mapping the spread of astroturf in microblog streams. In: Proceedings of the 20th International Conference Companion on World Wide Web, pp. 249–252. ACM (2011)
15. Rattanaritnont, G., Toyoda, M., Kitsuregawa, M.: A study on characteristics of topicspecific information cascade in Twitter. In: Forum on Data Engineering, DE 2011, pp. 65–70 (2011)
16. Romero, D.M., Meeder, B., Kleinberg, J.: Differences in the mechanics of information diffusion across topics: idioms, political hashtags, and complex contagion on Twitter. In: Proceedings of the 20th International Conference on World Wide Web, pp. 695–704. ACM (2011)
17. Sadikov, E., Martinez, M.M.M.: Information propagation on Twitter. CS322 Project Report (2009)
18. Sakaki, T., Okazaki, M., Matsuo, Y.: Earthquake shakes Twitter users: real-time event detection by social sensors. In: Proceedings of the 19th International Conference on World Wide Web, pp. 851–860. ACM (2010)
19. Watts, D.J., Strogatz, S.H.: Collective dynamics of 'small-world' networks. Nature **393**(6684), 440–442 (1998)
20. Zhou, Z., Bandari, R., Kong, J., Qian, H., Roychowdhury, V.: Information resonance on Twitter: watching iran. In: Proceedings of the First Workshop on Social Media Analytics, pp. 123–131. ACM (2010)
21. Zubiaga, A., Spina, D., Fresno, V., Martínez, R.: Classifying trending topics: a typology of conversation triggers on Twitter. In: Proceedings of the 20th ACM International Conference on Information and Knowledge Management, pp. 2461–2464. ACM (2011)

LiveRank: How to Refresh Old Crawls

The Dang Huynh[1,2], Fabien Mathieu[1(✉)], and Laurent Viennot[2]

[1] Alcatel-Lucent Bell Labs, Paris, France
`fabien.mathieu@alcatel-lucent.com`
[2] Inria – Univ. Paris Diderot, Paris, France

Abstract. This paper considers the problem of refreshing a crawl. More precisely, given a collection of Web pages (with hyperlinks) gathered at some time, we want to identify a significant fraction of these pages that still exist at present time. Liveness of an old page can be tested through an online query at present time. We call LiveRank a ranking of the old pages that tries to give good rankings to active nodes. The quality of a LiveRank is measured by the number of queries necessary to identify a given fraction of the alive pages when using the LiveRank order. We study different scenarios from a static setting where the LiveRank is computed before any query is made, to dynamic settings where the LiveRank can be updated as queries are processed. Our results show that building on the PageRank can lead to efficient LiveRanks for Web graphs.

1 Introduction

One of the main challenges for large networks data mining is to deal with the high dynamics of huge datasets: not only are these datasets difficult to gather, but they tend to become obsolete very quickly.

In this paper, we are interested in the evolution of large Web graphs at large time scale. We focus on batch crawling, where starting from a completely outdated snapshot of a large Web crawl, we want to identify a significant fraction of the pages that are still alive now.

Our motivation is that many old large snapshots of the Web are available today. Reconstructing roughly what remains from such archives could result in interesting studies of the long term evolution of these graphs. For large archives where one is interested in a fraction of the dataset, recrawling the full set of pages can be prohibitive. We propose to identify as quickly as possible a significant fraction of the pages that are still alive. Further selection can then be made to identify a set of pages suitable for the study and then to crawl them. Such techniques would be especially interesting when testing the liveness of an item is much lighter than downloading it completely. This is for instance the case for the Web with HEAD queries compared to GET queries. If a large amount of work has been devoted to maintaining fresh a set of crawled pages, little attention has been paid to the coverage obtained by partial recrawling a fairly old snapshot.

Problem formulation: Given an old snapshot, our goal is to identify a significant fraction of the pages that are still alive or active now. The cost we incur

© Springer International Publishing Switzerland 2014
A. Bonato et al. (Eds.): WAW 2014, LNCS 8882, pp. 148–160, 2014.
DOI: 10.1007/978-3-319-13123-8_12

is the number of fetches that are necessary to attain this goal. A typical cost measure will be the average number of fetches per active item identified. The strategy for achieving this goal consists in producing an ordering for fetching the pages. We call *LiveRank* an ordering such that the pages that are still alive tend to appear first. We consider the problem of finding an efficient LiveRank in three settings: static when it is computed solely from the snapshot and the link relations recorded at that time; sampling-based when a sampling is performed in a first phase allowing to adjust the ordering according to the liveness of sampled items; dynamic when it is incrementally computed as pages are fetched.

Contribution: We propose various LiveRank algorithms based on the graph structure of the snapshot. We evaluate them on two Web snapshots (from 10 to 20 million nodes). We show that a rather simple combination of a small sampling phase and PageRank-like propagation in the remaining of the snapshot allows to gather from 15% to 75% of the active nodes with a cost that remains within a factor of 2 from the optimal ideal solution.

Related work: The process of crawling the Web has been extensively studied. A survey is given by Olston and Najork [12].

The issue we investigate here is close to a problem introduced by Cho and Ntoulas [6]: they use sampling to estimate the frequency of change per site and then to fetch a set of pages such that the overall change ratio of the set is maximized. Their technique consists in estimating the frequency of page change per site and to crawl first sites with high frequency change. Tan et al. [15] improve slightly over this technique by clusterizing the pages according to several features: not only their site (and other features read from the URL) but also content based features and linkage features (including pagerank and incoming degree). A change ratio per cluster is then estimated through sampling and clusters are downloaded in descending order of the estimated values. More recently, Radinsky and Bennett [14] investigate a similar approach using learning techniques and avoiding the use of sampling.

Note that these approaches mainly focus on highly dynamic pages and use various information about pages whereas we are interested in stable pages and we use only the graph structure, which is lighter.

With a slightly different objective, Dasgupta et al. [7] investigate how to discover new pages while minimizing the average number of fetches per new page found. Their work advocates for: a greedy cover heuristic when a small fraction of the new pages has to be discovered quickly: an out-degree-based heuristic gathering a large fraction of the new pages. Their framework is close to ours and inspired the cost function used in this paper.

A related problem consist in estimating which pages are really valid among the "dangling" pages on the frontier of the crawled web (those that are pointed by crawled pages but that were not crawled themselves). Eiron et al. propose to take this into account in the PageRank computation [8]. In a similar trend, Bar-Yossef et al. [2] propose to compute a "decay" score for each page by refining on the proportion of dead links in a page. Their goal is to identify poorly updated pages. This score could be an interesting measure for computing a LiveRank, however its computation requires to identify dead links. It is thus not clear how to both estimate it and at the same time try to avoid to test dead pages.

Roadmap: In the next Section, we propose a simple cost function to evaluate the quality of a LiveRank and we introduce several classes of possible LiveRanks. In Section 3, we introduce two datasets from the .uk Web, for which we derived some ground truth of page liveness. Lastly, we benchmark our LiveRanks against these datasets and discuss the results in Section 4.

2   Model

Let $G = (V, E)$ be a directed graph obtained from a past Web snapshot, where V represents the crawled pages and E the hyperlinks: For i, j in V, (i, j) is in E if, and only if, there is a hyperlink to j in i.

Let n denote the size of V. At present time, only a subset of G are still alive (the exact meaning of liveness will be detailed in the next Section). We call a the function that tells if pages are alive or not: $a(X)$ denotes the alive pages from $X \subset V$, while $\bar{a}(X)$ stands for $X \setminus a(X)$. Let n_a be $|a(V)|$.

The problem we need to solve can be expressed as: how to crawl a maximum number of pages from $a(V)$ with a minimal crawling cost. In particular, one would like to avoid crawling too many pages from $\bar{a}(V)$. If a was known, the task would be easy, but testing the activity of a node obviously requires to crawl it. This is the rationale for the notion of LiveRank.

2.1   Performance Metric

Formally, any ordering on V can be seen as a LiveRank, so we need some performance metrics to measure the efficiency in ranking the pages from $a(V)$ first. Following [7], we define the LiveRank cost as the average number of page retrievals necessary to obtain one alive page, after a fraction $0 < \alpha \leq 1$ of the alive pages has been retrieved.

In details, let $\mathcal{L}_i$ represent the i first pages returned by a LiveRank $\mathcal{L}$, and let $i(\mathcal{L}, \alpha)$ be the smallest integer such that $\frac{|a(\mathcal{L}_i)|}{n_a} \geq \alpha$. The cost function of $\mathcal{L}$ (which depends on α) is then defined by:

$$\text{cost}(\mathcal{L}, \alpha) = \frac{i(\mathcal{L}, \alpha)}{\alpha n_a}.$$

A few remarks on the cost function:

- It is always at greater than or equal to 1. An ideal LiveRank would perfectly separate $a(V)$ from rest of the nodes, so its cost function would be 1. Without some oracle, this requires to test all pages, which is exactly what we would like to avoid. The cost function allows to capture this dilemma.
- Keeping a low cost becomes hard as α gets close to 1: without some clairvoyant knowledge, capturing *almost* all active nodes is almost as difficult as capturing all actives nodes. For that reason, one expects that when α gets close to 1, the set of nodes any real LiveRank will need to crawl will tend to V, leading to an asymptotical cost $\frac{n}{n_a}$. This will be verified in Section 4.

- Lastly, one may have noticed that the cost function uses $n_a = |a(V)|$, for which an exact value requires a full knowledge of liveness. This is not an issue here as we will perform our evaluation on datasets where a is known. For use on datasets without ground truth, one could either use an estimation of n_a based on a sampling or use a non-normalized cost function (for instance the fraction of alive pages obtained after i retrievals).

2.2 PageRank

Some of the proposed LiveRanks are based on PageRank. PageRank is a link analysis algorithm introduced in [13] and used by the Google Internet search engine. It assigns a numerical importance to each page of a Web graph. It uses the structural information from G to attribute importance according to the following (informal) recursive definition: *a page is important if it is referenced by important pages*. Concretely, to compute PageRank value, denoted by the row vector Y, one needs to find the solution of the following equation:

$$Y = dYA + (1 - d)X, \tag{1}$$

where A is a substochastic matrix derived from the adjacency matrix of G, $d < 1$ a so-called damping factor (often set empirically to $d = 0.85$), and $X \gneqq 0$ is a *teleportation vector*. X represents a kind of importance *by default* that is propagated from pages to pages according to A with a damping d.

Computation of PageRank vectors has being widely studied. Several specific solutions were proposed and analysed [3,11] including power method [13], extrapolation [9,10], adaptive on-line method [1], etc.

We now present the different LiveRanks that we will consider in this paper. We broadly classify them in three classes: static, sample-based and dynamic.

2.3 Static LiveRanks

Static LiveRanks are computed offline using uniquely the information from G. That makes them very basic, but also very easy to be used in a distributed way: given p crawlers of similar capacities, if $\mathcal{L} = (l_1, \ldots, l_n)$, simply assign the task of testing node l_i to crawler $i \mod p$.

We propose the following three static LiveRanks.

Random Permutation (R). will serve both as a reference and as a building block for more advanced LiveRanks. R ignores any information from G, so its cost should be in average $\frac{n}{n_a}$, with a variance that tends to 0 as α tends to 1. We expect good LiveRanks to have a cost function significantly lower than cost(R).

Decreasing Indegree Ordering (I). is a simple LiveRank that we expect to behave better than a random permutation. Intuitively, a high Indegree can mean some importance, and important pages may be more robust. Also, older pages should have more incoming edges (in terms of correlation), so high degree pages can correspond to pages that were already old at the time G was crawled, and old pages may last longer than younger ones. Sorting by degree is the easiest way to exploit these correlations.

PageRank Ordering (P). pushes forward the *indegree* idea. The intuition is that pages that are still alive are likely to point toward pages that are still alive also, even considering only old links. This suggests to use a PageRank-like importance ranking. In absence of further knowledge, we propose to use the solution of (1) using $d = .85$ (typical value for Web graphs) and X uniform on V.

Note that it is very subjective to evaluate PageRank as an importance ranking, as importance should be ultimately validated by humans. On the other hand, the quality of PageRank as a static LiveRank is straightforward to verify, for instance using our cost metric.

The possible existence of correlation between Indegree (or PageRank) and liveness will be verified in Section 3.3.

2.4   Sample-Based LiveRanks

Using a LiveRank consists in crawling V in the prescribed order. During the crawl, the activity function a becomes partly available, and the obtained information could be used to enhance the retrieval. Following that idea, we consider here a two-steps sample-based approach: we first fix a testing threshold z and test z items following a static LiveRank (like R, I or P). For the set Z of nodes tested, called *sample set* or *training set*, $a(Z)$ is known, which allows us to recompute the LiveRank of the remaining untested pages.

Because the sampling uses a static LiveRank, and the adjusted new LiveRank is static as well, sample-based LiveRanks are still easy to use in a distributed way as the crawlers only need to receive crawl instructions on two occasions.

Notice that in the case where the sampling LiveRank is a random permutation, $|a(Z)|\frac{n}{z}$ can be used as an estimate for n_a. This can for instance be used to decide when to stop crawling if we desire to identify αn_a active nodes in $a(V)$.

Simple Adaptive LiveRank (P_a). When a page is alive, we can assume it increases the chance that pages it points to in G are also alive, and that life is transmitted somehow through hyperlinks. Following this idea, a possible adaptive LiveRank consists in taking for X in (1) the uniform distribution on $a(Z)$. This diffusion from such an initial set can be seen as a kind of breadth-first traversal starting from $a(Z)$, but with a PageRank flavour.

Double Adaptive LiveRank $(P_a^{+/-})$. The simple adaptive LiveRank does not use the information given by $\bar{a}(Z)$. One way to do this is to calculate an "anti"-PageRank based on $\bar{a}(Z)$ instead of $a(Z)$. This ranking would represents a kind of diffusion of death, the underlying hypothesis being that dead pages may point to pages that tend to be dead. As a result, we obtain a new LiveRank by combining these two PageRanks. After having tested several possible combinations not discussed in this paper, we empirically chose to weight each node by the ratio of the two sample-based PageRank, after having set all null entries of the anti-PageRank equal to the minimal non-null entry.

Active-Site First LiveRank (ASF). To compare with previous work, we propose the following variant inspired by the Dasgupta et al. [7] strategy for finding pages that have changed in a recrawl. Their algorithm is based on sampling for estimating page change rate for each website and then to crawl sites by decreasing change rate. In details, Active-site first (ASF) consists in partitioning Z into websites determined by inspecting the URLs. We thus obtain a collection $Z_1, \ldots, Z_p$ of sets. For each set Z_i corresponding to some site i, we obtain an estimation $|a(Z_i)|/|Z_i|$ of its activity (i.e. the fraction of active pages in the site). We then sort the remaining URLs by decreasing site activity.

2.5   Dynamic LiveRanks

Instead of using the acquired information just one time after the sampling, Dynamic LiveRanks are continuously computed and updated on the fly along the entire crawling process. On the one hand, this gives them real-time knowledge of a, but on the other hand, as the dynamic LiveRank may evolve all the time, they can create synchronization issues when used by distributed crawlers.

Like for sample-based LiveRanks, dynamic LiveRanks use a training set Z of z pages from a static LiveRank. This allows to bootstrap the adjustment by giving a non-empty knowledge of a, and prevents the LiveRank from focusing on only a small subset of V.

Breadth-First Search (BFS). With BFS, we aim at taking direct advantage of the possible propagation of liveness. The BFS queue is initialized with the (uncrawled) training set Z. The next page to be crawled is popped from the queue following First-In-First-Out (FIFO) rule. If the selected page appears to be alive, all of its uncrawled outgoing neighbors are pushed into the end of the queue. When the queue is empty, we pick the unvisited page with highest PageRank[1].

Alive Indegree (AI). BFS uses a simple FIFO queuing to determine the processing order. We now propose AI which provides a more advanced page selection scheme. For AI, each page in the graph is associated with a *live score* value indicating how many reported alive pages point to it. These values are set to zeros at the beginning and always kept up-to-date. AI is initialized by testing Z: each node in $a(Z)$ will increment the associated values of its out-going neighbors by one. After Z is tested, the next node to be crawled is simply the one with highest live score (in case of equality, to keep things consistent, we pick the node with highest PageRank). Whenever a new alive node is found, we update the live scores of its untested neighbors.

With Dynamic LiveRank, it is natural to think of a dynamic PageRank-based strategy where PageRank vector is recursively computed. Starting from a uniform distribution on $a(Z)$, we obtain X in (1). Then a new teleportation vector is constructed as a uniform distribution on largest value entries of X,

---

[1] We tested several other natural options and observed no significant impact.

i.e., those which are considered probably alive after the first diffusion of $a(Z)$. The process continues and X is updated iteratively. However, this method is not efficient since it can not escape from the locality of $a(Z)$.

3   Datasets

We chose to evaluate the proposed LiveRanks on datasets of the British domain .uk available on the WebGraph platform[2]. In this Section, we present these datasets, describe how we obtained the alive function a and observe the correlations between a, indegree and PageRank.

3.1   uk-2002 Dataset

The main dataset we use is the web graph uk-2002[3] from UbiCrawler [4]. This 2002 snapshot contains 18,520,486 pages and 298,113,762 hyperlinks.

The preliminary task is to determine a, the liveness of the pages of the snapshot. For each URL, we have performed a GET request and hopefully obtained a corresponding HTTP code. Our main findings are:

– One third of the total pages are no longer available today, the server returns error 404.
– One fourth have a DNS problem (which probably means the website is also dead).
– For one fifth of the cases, the server sends back the redirection message 301. Most redirections for pages of an old site lead to the root of a new site. If we look at the proportion of distinct pages alive at the end of redirections, it is as low as 0.1%.
– Less than 13% of pages return the code 200 (success). However, we found out that half of them actually display some text mentioning that the page was not found. To handle this issue, we have fully crawled all the pages with code 200 and filtered out pages whose title or content have either Page Not Found or Error 404.

The results are summarized in Table 1. In the end, our methodology led to finding out 1,164,998 alive pages, accounting for 6.4% of the dataset.

3.2   uk-2006 Dataset

The settings of uk-2002 are rather adversarial (old snapshot with relatively few alive pages), so we wanted to evaluate the impact of LiveRanks on shorter time scales. In absence of fresh enough available datasets, we used the DELIS dataset [5], a series of twelve continuous snapshots[4] starting from 06/2006 to

---

[2] http://webgraph.di.unimi.it/
[3] http://law.di.unimi.it/webdata/uk-2002/
[4] http://law.di.unimi.it/webdata/uk-union-2006-06-2007-05/

Table 1. Status of web pages in uk-2002, crawled in December 2013

Status	Description	Number of pages	Percentage
Code HTTP 404	Page not found	6 467 219	34,92%
No answer	Host not found	4 470 845	24,14%
Code HTTP 301	Redirection	3 455 923	18,66%
Target 301	Target of redirection	20 414	0,11%
Code HTTP 200	Page exists	2 365 201	12,77%
True 200	**Page really exists**	1 164 998	**6,29%**
Others (403,...)	Other error	1 761 298	9,51%
Total	Graph size	18 520 486	100%

05/2007 (one-month intervals). We set G to the first snapshot (06/2006). It contains 31,316,403 nodes and 813,807,972 hyperlinks. We then considered the last snapshot (05/2007) as "present time", setting the active set $a(V)$ as the intersection between the two snapshots. With this methodology, we hope to have a good approximation of a after a one-year period. For this dataset, we obtained $n_a = 11,142,177$ "alive" nodes representing 35.56% of the graph.

3.3 Correlations

The rationale behind the LiveRanks I and P is the assumption that the liveness of pages is correlated to the graph structure of the snapshot, so that a page with high in-degree or PageRank has more chances to stay alive.

To validate this, we plot in Figure 1 the cumulative distribution of in-degree (figure 1a) and PageRank (figure 1b) for alive, dead, and all pages of the uk-2002 dataset. We observe that the curve for active nodes is slightly shifted to the right compared to the other curves in each figures: active users tend to have slightly higher in-degree and PageRank than in the overall population. The bias is bigger for PageRank, suggesting that LiveRank (P) should perform better than LiveRank (I).

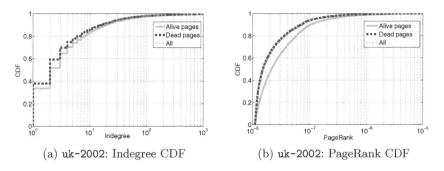

(a) uk-2002: Indegree CDF   (b) uk-2002: PageRank CDF

Fig. 1. Cumulative distribution of pages according to Indegree and PageRank

4   LiveRanks Evaluation

After having proposed several LiveRanks in Section 2 and described our datasets in previous Section, we can now benchmark our proposals.

All our evaluations are based on representations of the cost functions. In each plot, the x-axis indicates the fraction α of active nodes we aim to discover and the y-axis corresponds to the relative cost of the crawl required to achieve that goal. A low curve indicates an efficient LiveRank. Like said in Section 2.1: an ideal LiveRank would achieve a constant cost of 1; a random LiveRank is quickly constant with an average cost n/n_a; any non-clairvoyant LiveRank will tend to cost n/n_a as α goes to 1.

We mainly focus on the uk-2002 dataset. When it is not specified, the training set contains the $z = 100000$ pages of higher (static) PageRank.

4.1   Static and Sample-Based LiveRanks

We first evaluate the results of static and sample-based LiveRanks. The results are displayed in Figure 2. For static LiveRanks, we see as expected that a random ordering gives an almost constant cost equal to $\frac{n}{n_a} \approx 15.6$. Indegree ordering (I) and PageRank (P) significantly outperform this result, PageRank being the best of the three: it is twice more efficient than random for small α, and still performs approximately 30% better when up to $\alpha = 0.6$. We then notice that we can get even much better costs with sample-based approaches, the double-adaptive LiveRank $P_a^{+/-}$ giving a significant improvement over the simple-adaptive one P_a. $P_a^{+/-}$ allows improving the ordering by a factor of 6 approximately around $\alpha = 0.2$ with a cost of 2.5 fetches per active node found. The cost for gathering half of the alive pages is less than 4, and for 90% it stays less than 10.

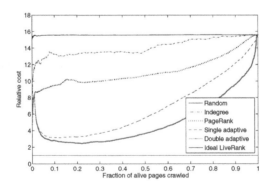

Fig. 2. Main results on static and sample-based LiveRanks

4.2   Quantitative and Qualitative Impact of the Training Set

We study in Figure 3 the impact of the training sets on sample-based LiveRanks. Results are shown for $P_a^{+/-}$ but similar results were obtained for P_a.

Figure 3a shows the impact of the size z of the sampling set (sampling the top PageRank pages). We observe some trade-off: as the sampling set grows larger, the initial cost increases as the sample does not used any fresh information, but it results in a significant increment of efficiency in the long run. For this dataset, taking a big training set (z=500 000) allows reducing the cost of the crawl for $\alpha \geq 0.4$, and maintains a cost less than 4 for up to 90%.

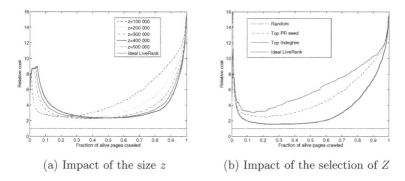

(a) Impact of the size z (b) Impact of the selection of Z

Fig. 3. Impact of the training set

Another key aspect of the sampling phase is the qualitative choice of the sample set. Using z=100 000, we can observe in Figure 3b that the performance of double adaptive $P_a^{+/-}$ is further improved by using a random sample set rather than selecting it according to the PageRank or by decreasing indegree. We believe that the reason is that a random sample avoids a locality effect in the sampling set as high PageRank pages tend to concentrate in some local parts of the graph. To verify that, we tried to modify Indegree and PageRank selection to avoid to select neighbor pages. The results (not displayed here) show a significant improvement while staying less efficient than using a random sample.

Note that double-adaptive LiveRank through random sampling offers a very low cost, within a factor of 2 from optimal for a large range of values α.

4.3   Dynamic LiveRanks

We then consider the performance of fully dynamic strategies, using the double-adaptive LiveRank with random training set as a landmark. The results are displayed in Figure 4a. We see that bread-first search BFS and alive indegree AI perform similarly to double adaptive $P_a^{+/-}$ for low α and can outperform it for large α (especially BFS). BFS begin to significantly outperform double adaptive for $\alpha \geq 0.5$. However, if one needs to gather half of the active pages or less, double adaptive is still the best candidate as it is much simpler to operate, especially with a distributed crawler.

Additionally, Figure 4b shows the impact of different sampling sets on BFS and AI. Except for high values of α where a random sampling outperforms other strategies, the type of sampling does not seem to affect the two dynamic LiveRanks as much as it was observed for the double-adaptive LiveRank.

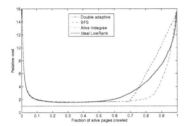
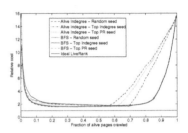

(a) Performance of dynamic LiveRanks (b) Impact of Z on dynamic LiveRanks

Fig. 4. uk-2002 Performance of dynamic LiveRanks

4.4   uk-2006 Dataset

We have repeated the same experiments on the dataset uk-2006, where the update interval is only one year. Figure 5 shows the results for static and sample-based LiveRanks, using z=200 000 (because the dataset is larger) and random sampling. The observation are qualitatively quite similar to uk-2002. The main difference is that all costs are lower due to a higher proportion of alive pages ($\frac{n}{n_a} \approx 2.81$). The double-adaptive version still gives the lower relative cost among static and sample-based LiveRanks, staying under 1.4 for a wide range of α.

4.5   Comparison with a Site-Based Approach

To benchmark with techniques from previous work for finding web pages that been updated after a crawl, Figure 6 compares double adaptive $P_a^{+/-}$ to active-site first ASF with random sampling. The number of random pages tested in each site and the overall number of tests are the same for both methods. Note that given the budget z, it was not possible to sample small websites. Unsampled websites are crawled after the sampled ones.

We see that for α greater than 0.9, ASF performs like a random LiveRank. This corresponds to the point where all sampled website have been crawled. That effect aside, the performance of ASF is not as good as double-adaptive LiveRank for earlier α. In the end, ASF only beats $P_a^{+/-}$ for a small range of α, between 0.7 and 0.85, and the gain within that range stays limited.

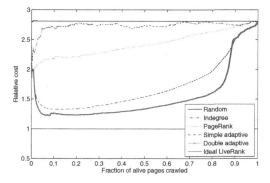

Fig. 5. uk-2006 main evaluation results

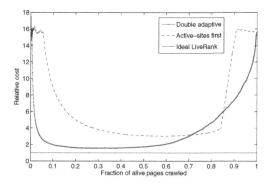

Fig. 6. Comparison with the cost of an active-site first LiveRank

5 Conclusion

In this paper, we investigated how to efficiently retrieve large portions of alive pages from an old crawl using orderings we called LiveRanks. We observed that PageRank is a good static LiveRank, which can be significantly improved by first testing a small fraction of the pages for adjustment in a sample-based approach.

Compared to previous work on identifying modified pages, our technique performs similarly for a given large desired fraction (around 80%) when compared to the LiveRank algorithm inspired by the technique in [6]. However, outside that range, our method outperforms this technique. Interesting future work could reside in using our techniques for the problem exposed in [6] (identification of pages that have changed) and compare with the Website sampling approach.

Interestingly, we could not get significant gain when using fully dynamic LiveRanks. As noticed before, each of the two phases of the sample-based approach can be easily parallelized through multiple crawlers whereas this would be much more difficult with a fully dynamic approach. The sample-based method could for example be implemented with in two rounds of a simple map-reduce program whereas the dynamic approach requires continuous exchanges of messages between the crawlers.

Our work establishes the possibility of efficiently recovering a significant portion of the alive pages of an old snapshot and advocates for the use of an adaptive sample-based PageRank for obtaining an efficient LiveRank.

To conclude, we emphasize that the LiveRank approach proposed in this paper is very generic, and its field of applications is not limited to Web graphs. It can be straightforwardly adapted to any online data with similar linkage enabling crawling, like P2P networks or online social networks. For future work, our approach will be mathematically extended. The problem can be formulated as an accuracy estimation of LiveRank vector, given a uniform distribution on the alive training set as teleportation vector.

References

1. Abiteboul, S., Preda, M., Cobena, G.: Adaptive on-line page importance computation. In: WWW 2003, pp. 280–290. ACM (2003)
2. Bar-Yossef, Z., Broder, A.Z., Kumar, R., Tomkins, A.: Sic transit gloria telae: Towards an understanding of the web's decay. In: WWW 2004, pp. 328–337 (2004)
3. Bianchini, M., Gori, M., Scarselli, F.: Inside pagerank. ACM Trans. Internet Technol. 5(1), 92–128 (2005)
4. Boldi, P., Codenotti, B., Santini, M., Vigna, S.: Ubicrawler: A scalable fully distributed web crawler. Software: Practice & Experience 34(8), 711–726 (2004)
5. Boldi, P., Santini, M., Vigna, S.: A large time-aware graph. SIGIR Forum 42(2), 33–38 (2008)
6. Cho, J., Ntoulas, A.: Effective change detection using sampling. In: VLDB 2002, pp. 514–525 (2002)
7. Dasgupta, A., Ghosh, A., Kumar, R., Olston, C., Pandey, S., Tomkins, A.: The discoverability of the web. In: WWW 2007, pp. 421–430. ACM (2007)
8. Eiron, N., McCurley, K.S., Tomlin, J.A.: Ranking the web frontier. In: WWW 2004, pp. 309–318. ACM (2004)
9. Haveliwala, T., Kamvar, A., Klein, D., Manning, C., Golub, G.: Computing pagerank using power extrapolation, Technical report (2003)
10. Kamvar, S.D., Haveliwala, T.H., Manning, C.D., Golub, G.H.: Extrapolation methods for accelerating pagerank computations. In: WWW 2003, pp. 261–270. ACM (2003)
11. Langville, A.N., Meyer, C.D.: Deeper inside pagerank. Internet Mathematics 1 (2004)
12. Olston, C., Najork, M.: Web crawling. Foundations and Trends in Information Retrieval 4(3), 175–246 (2010)
13. Page, L., Brin, S., Motwani, R., Winograd, T.: In: The PageRank Citation Ranking: Bringing Order to the Web., number 1999–66. Stanford InfoLab (1999)
14. Page, L., Brin, S., Motwani, R., Winograd, T.: The PageRank Citation Ranking: Bringing Order to the Web., number 1999–66. Stanford InfoLab (1999)
15. Tan, Q., Zhuang, Z., Mitra, P., Giles, C.L.: A clustering-based sampling approach for refreshing search engine's database. In: WebDB 2007 (2007)

Author Index

Avrachenkov, Konstantin 23

Bahmani, Bahman 59
Bhulai, Sandjai 132
Blesa, Maria J. 108
Boehnlein, Edward 79
Bonato, Anthony 13

Candellero, Elisabetta 1
Chen, Ningyuan 120
Chin, Peter 79

Fountoulakis, Nikolaos 1

Goel, Ashish 59

Huynh, The Dang 148

Janssen, Jeannette 13

Lattanzi, Silvio 34
Leonardi, Stefano 34
Li, Jiankou 96

Litvak, Nelly 120, 132
Lu, Linyuan 79

Mathieu, Fabien 148
Molter, Hendrik 108
Munagala, Kamesh 59

Olvera-Cravioto, Mariana 120
Ostroumova Prokhorenkova, Liudmila 47
Ouboter, Tanneke 132

Roshanbin, Elham 13

Samosvat, Egor 47
Sinha, Amit 79
Sokol, Marina 23

Viennot, Laurent 148

Worm, Daniël 132

Yong, Xi 96

Zhang, Wei 96

Printed in the United States
By Bookmasters